Frank Rutley

The Eruptive Rocks of Brent Tor and its Neigbourhood

Included in sheet 25 of the one-inch map of the Geological survey, with

some introductory remarks on the application of the microscope to

petrological research

Frank Rutley

The Eruptive Rocks of Brent Tor and its Neigbourhood
Included in sheet 25 of the one-inch map of the Geological survey, with some introductory remarks on the application of the microscope to petrological research

ISBN/EAN: 9783337377366

Printed in Europe, USA, Canada, Australia, Japan

Cover: Foto ©berggeist007 / pixelio.de

More available books at **www.hansebooks.com**

MEMOIRS OF THE GEOLOGICAL SURVEY.

ENGLAND AND WALES.

THE

ERUPTIVE ROCKS

OF

BRENT TOR AND ITS NEIGHBOURHOOD,

INCLUDED IN

SHEET 25 OF THE ONE-INCH MAP OF THE GEOLOGICAL SURVEY;

WITH SOME

INTRODUCTORY REMARKS ON THE APPLICATION OF THE MICROSCOPE TO PETROLOGICAL RESEARCH.

BY

FRANK RUTLEY, F.G.S.

PUBLISHED BY ORDER OF THE LORDS COMMISSIONERS OF HER MAJESTY'S TREASURY.

LONDON:
PRINTED FOR HER MAJESTY'S STATIONERY OFFICE,
AND SOLD BY
LONGMAN & Co., Paternoster Row; TRÜBNER & Co., Ludgate Hill;
LETTS & SON, 33, King William Street;
EDWARD STANFORD, 55, Charing Cross; and J. WYLD, 12, Charing Cross:
ALSO BY
Messrs. JOHNSTON, 4, St. Andrew Square, Edinburgh;
HODGES, FOSTER, & Co., 104, Grafton Street, and A. THOM,
Abbey Street, Dublin.

1878.

Price Fifteen Shillings and Sixpence.

THE following Memoir by Mr. Rutley is an important contribution to the history of Brent Tor, and the surrounding igneous and metamorphosed rocks of that part of Devonshire. His remarks on the microscopic analyses of the minerals which compose the rocks, and the beauty of the coloured plates prepared by the author, will be appreciated by all who are interested in such subjects.

It is not uncommon in modern memoirs on petrographical subjects, for authors to criticise the nomenclatures of old authors, as if they had been possessed of all the knowledge that the microscope, polariscope, and analytical chemistry bring to bear on petrology.

If on our national Geological Maps all the mineralogical variations in the constitution of igneous and metamorphic rocks were to be classified on such principles, the number of colours and shades of colours available would probably be insufficient to compass all of the stratigraphical and petrological distinctions which are and will be made—the number at present employed on the Geological Survey being 103. It is therefore satisfactory to those engaged in the broadly geological work of mapping to find, as Mr. Rutley remarks, " that the " views entertained by the late Sir Henry De la Beche " are in the main correct, and represent a vast amount " of truth derived simply from observation in the field." This opinion is also entertained by Mr. J. A. Phillips, in a late memoir in the Journal of the Geological Society, in which he states that notwithstanding mineralogical differences in many of the igneous rocks of Devon and

Cornwall, the fact that their ultimate analyses is almost the same, forms a new and additional justification for Sir Henry De la Beche having often adopted one colour and one general name for rocks, the mineral peculiarities of which are, indeed, sometimes only due to subsequent metamorphic action.

AND. C. RAMSAY,
Director-General.

Geological Survey Office,
28, Jermyn Street, S.W.
27th June 1878.

NOTICE.

This, the first special petrographical work issued by the Geological Survey, has been written by Mr. Rutley after a personal inspection of the rocks in the field, followed by a microscopical examination of specimens carefully selected by himself.

The district treated of was surveyed nearly half a century ago, by Sir Henry De la Beche, who recorded his observations and deductions in his report on the geology of Cornwall, Devon, and West Somerset, published in the year 1839, to which work Mr. Rutley's investigations may be regarded as a sequel.

At that time the microscope (except in the form of pocket-lens), was little used in geological research, and it is only comparatively lately that the importance of determining the mineral composition of eruptive rocks, and of accurately investigating their mineral structure, has become fully recognized, as the light which such investigations throw upon conditions under which those rocks were formed has become more and more apparent.

·In the present case the old geological boundary-lines, as originally surveyed by Sir Henry De la Beche, and now engraved upon the maps of the Geological Survey, hold good as a rule, or, if defective, only require very slight modication.

The present Memoir will, I believe, prove an important and valuable addition to the other publications of the Geological Survey, from the beauty and fidelity of Mr. Rutley's microscopic drawings, as well as from the clear and masterly way in which he has treated the subject.

The eruptive rocks, which are now collected during the progress of the Survey in mapping new districts, are subjected to microscopic examination, where such examination seems needful—prior to the publication of the maps and memoirs on the districts in which they occur—

C. 31.   A and the specimens thus examined will eventually be placed in the Collection of Rocks when re-arranged in the new room which will shortly be assigned to it in the Museum of Practical Geology. The results of these examinations will also be embodied in future editions of the printed Catalogue of Rock Specimens, which will henceforth have additional scientific as well as practical interest in its bearings on a right interpretation of the geology of the British Isles.

H. W. Bristow,
Senior Director.

Geological Survey Office,
28, Jermyn Street, S.W.
19th June 1878.

PREFACE.

SINCE the days of Sir Henry De la Beche, little has been done
to elucidate the geological relations and the mineral composition

TABLE OF CONTENTS.

LIST OF PLATES.

CHROMOLITHOGRAPHS OF MICROSCOPIC SECTIONS.

LIST OF WOODCUTS.

INTRODUCTION.

THE present Memoir being the first special petrographical study hitherto published by the Geological Survey, it may be well to make a few remarks upon the use of the microscope in this branch of geological research.

Until within the last few years the minute structure of rocks was comparatively unknown, and very great uncertainty existed as to the mineral composition of many, while that of the fine-grained varieties was a matter of mere speculation.

The pocket lens in the hands of the older geologists was the sole means which they possessed of gaining an insight into the mineral composition and structure of rocks which presented no definite character to the unaided eye ; for although chemistry afforded them a rough idea of the minerals which a rock might contain, still it failed to indicate definite mineral species; and structural peculiarities too minute to be recognised by the pocket lens remained unknown. In spite, however, of the advance which has been made of late years in this branch of petrological knowledge by the help of the compound microscope, it must not be supposed that the pocket lens is a comparatively useless instrument. Much may be done by its aid, and often with better results than would be attained by the use of higher magnifying powers. This may be partly attributed to the greater ease with which we recognise minerals when seen by reflected light, and when showing actual crystalline faces and planes of cleavage, than when seen by transmitted light and presenting usually only the boundaries of crystals, and the little vestiges of cleavage which remain between the abraded planes which limit a thin microscopic section.

The compound microscope is undoubtedly a most valuable aid in determining many of the minerals which compose rocks, but it should always be borne in mind that when the component minerals of a rock are sufficiently well developed to be recognised under a pocket lens by their crystalline form, cleavage, lustre, or by other tests, such as hardness, colour of streak, or blowpipe characters, it is better to rely more upon such evidence than upon the appearances seen under high magnifying power, especially when the latter do not happen to be very characteristic. It is, however, often impossible to arrive at any sound conclusions respecting the mineral composition of some rocks by the simple method just indicated. Thus, for example, in the Obsidians and

Pitchstones many interesting structures occur, which in some
instances serve to indicate points connected with the solidification
of those rocks, while in other cases microliths of augite, &c.,
crystals of felspars and micas are seen to constitute by no means
an unimportant part of some of them.   In fine-grained aphanitic
rocks also, microscopic examination is the only method by which
we can form any definite notion of the different mineral species
of which they are composed, and even in rocks whose coarsely
crystalline structure permits a ready determination of many of
their components, the microscope often reveals the presence of
large numbers of minute crystals of other minerals, the existence
of which would otherwise have eluded notice, or have rendered
analytical results difficult of explanation.

The employment of polarized light in microscopic examination
is also of very great assistance in determining the components of
rocks.   In many cases the crystalline systems to which felspars
may be referred may be ascertained with considerable certainty,
those belonging to the triclinic system exhibiting parallel bands
when viewed by polarized light, the bands being due to the con-
tiguous development of twin lamellæ.   These bands are often very
numerous in plagioclase crystals, but unless the section examined
be cut at right angles to the planes of composition, the overlap of
adjacent lamellæ may give rise to colourless stripes, as shown in
Fig. 1, where A and B represent twin lamellæ polarizing in com-
plentary colours, the colourless bands produced by the overlaps of
the different lamellæ being indicated by the unshaded stripes in
the lower figure, and an estimate of the number of lamellæ by
the number of stripes would be liable to error.

Fig. 1.

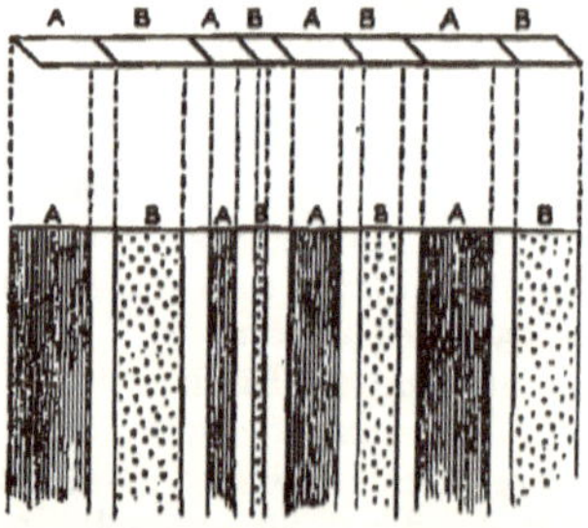

In the Orthoclastic felspars the twinning usually takes place
upon what is known as the " Carlsbad type," viz., in a plane
parallel with the clino-pinakoid, and I believe that no instances
are known in which an orthoclastic felspar exhibits more than
two such lamellæ in an individual crystal.   Orthoclase crystals are
also twinned at times upon another plan known as the " Baveno
type."   In this case, the twinning plane extends between the
alternate edges formed by the basis and the clino-pinakoid, but
microscopic orthoclase crystals twinned upon this type are seldom

or never seen. In some examples of massive Orthoclase, and occasionally, but rarely in crystals, a very peculiar cross-hatched structure is shown by polarized light. The true nature of this marking is not yet satisfactorily ascertained, but it is very characteristic of Orthoclase.

By means of the microscope many felspars are seen to be anything but homogeneous; patches, and, as in the case of Perthite, lamellæ of one species being intermixed with another. Not merely does the microscope demonstrate the presence of impurities in felspars, but in many, if not in most, other minerals. It is also interesting to note, by its means, the way in which alteration products or minerals of secondary origin are distributed in rocks; how augite and hornblende crystals have been replaced by serpentinous matter; how at times an augite crystal includes little patches of hornblende, which shade away into the surrounding mineral, and apparently represent an incipient stage of the development of uralite. The alterations which ferruginous minerals undergo are also well shown at times in microscopic sections. Magnetite becomes altered into hydrous peroxide of iron; the titaniferous iron minerals also show curious alterations, and pyrites and limonite are common as secondary products in many rocks. The examination of these minerals can only be effected, as a rule, by means of reflected light, since they are, in nearly all cases, too opaque to present any definite character by transmitted illumination, other than the outlines of crystals. These outlines enable us at times to discriminate between magnetite and titaniferous iron; but to distinguish magnetite from pyrites it is needful to employ reflected light, and even then a satisfactory determination can often only be effected by blowpipe or other analysis.

Felspars, both monoclinic and triclinic, frequently undergo a considerable amount of change, a granular condition supervening, which, if far advanced, obliterates all internal structure, so that the felspars belonging to the one system cannot be distinguished from those of the other. This change is generally attributed to weathering; but upon examining a section of basalt from Debdon near Rothbury, Northumberland, cut so as to display the pale grey weathered crust and the dark subjacent rock, I was much surprised to find that the plagioclase crystals presented just as fresh an appearance as those in the unweathered portion of the rock. If this be an exceptional instance I do not know how to account for it; certain it is, however, that in the majority of cases where the weathering of a rock is far advanced and minerals of secondary origin are plentifully developed, the felspars very frequently exhibit the clouded and granular texture just alluded to. This is especially the case with the felspars in diabase.

The method first pointed out by Tschermak of distinguishing between the minerals hornblende and augite by means of the dichroism of the former, and the absence of that character in the latter mineral, appears up to a certain point to suffice; still some augites, especially if the sections be thick, show slight dichroism when a Nicol's prism is rotated beneath them, while in some exceedingly thin sections of hornblende the dichroism is very feeble. The test, as a rule, however, seems to hold good, while when well-marked transverse sections of these minerals occur the great discrepancy between the angles of the oblique rhombic prism afford a ready means of distinguishing between them. The dichroism shown by hypersthene and enstatite, and the absence or almost total absence of that character in diallage and bronzite, also serve to some extent to facilitate the recognition of these minerals.

Considerable difficulty may, however, be experienced at times in distinguishing between hornblende and magnesian mica when these minerals are disseminated through a section in excessively minute flecks, and when no definite crystalline form or cleavage is discernible, the colours of the two minerals being often much alike, and their dichroism when the mica is cut at right angles or obliquely to the basal plane being of about equal intensity. Under these circumstances it is well to search carefully over hand specimens of the rock with a strong magnifier, in the hope of finding a moderately well-developed crystal.

The characters by which some minerals are recognised under the microscope are comparatively simple. Olivine, for example, may generally be known by the roughened appearance which the ground surfaces of its sections present, and by the rounded angles usually shown when the crystals are definitely formed. Its want of dichroism, its comparatively weak chromatic polarization, and its freedom from inclosures of other minerals also serve to distinguish it. When unaltered, the crystals appear homogeneous in composition and structure, but when they have undergone change the result is usually serpentinous matter, which under polarized light breaks up into feebly polarizing and irregular patches. Schorl may usually be distinguished from hornblende by the trigonal transverse sections which sometimes occur in a thin slice of rock, and by the bluish tint which it frequently exhibits when seen by ordinary transmitted light, especially when it occurs associated with quartz. The hemihedral terminations of crystals are not often seen in rock sections, and consequently this characteristic is not, as a rule, to be observed.

Nepheline and apatite may be recognised by their hexagonal transverse sections, and by the total extinction of light which takes place in such sections between crossed nicols. The former mineral usually occurs in larger crystals than the latter, the apatite crystals frequently being developed upon a very small scale, so that they appear as mere spiculæ or microliths under tolerably high magnifying power. Crystals of apatite often form small colonies in certain spots in a rock, their distribution being by no means regular. Apatite is a mineral of very common occurrence in British eruptive rocks, but as yet the only known instance of the occurrence of nepheline is in the Wolf Rock, off the coast of Cornwall.* The longitudinal sections both of nephe-

---

* *Vide* paper by S. Allport, F.G.S., Geol. Mag., Vol. VIII., page 247.

line and apatite are rectangular in form, and polarize in more or less vivid colours when the crystals are of moderate size.

Leucite is a mineral as yet unknown in British rocks. If, however, it should occur developed only in very minute crystals it might easily be overlooked or mistaken for some other mineral, since very small crystals of leucite, such as those which sometimes occur in the leucitophyrs near Rome, do not display any of the characteristics of the larger crystals, and their boundaries are often so ill-defined that they look merely like rounded singly refracting granules.

In the absence of definite crystals, or of characteristic crystalline aggregates, it is often difficult to distinguish between many of the hydrous alkaline silicates, which occur so plentifully as secondary products in many rocks which have reached different stages of decomposition. These mostly present a green colour when seen under the microscope by ordinary transmitted light, and, owing to the difficulty in referring them to definite species, Vogelsang proposed the use of the term viridite by which to designate them, until further study should enable the observer to assign them with precision to their respective species. The scaly forms of viridite are usually considered to be some variety of chlorite, while the fibrous kinds are regarded as serpentinous matter. Opacite and ferrite are also provisional terms; the former applies to a black, opaque, amorphous substance, which often occurs pseudomorphous after other minerals. It may represent either earthy silicates or amorphous metallic oxides. The latter term is given to earthy matter, probably oxides of iron, hydrous and anhydrous.

From the foregoing remarks it will be seen that at the present time there are several terms in use in mico-petrology which have a very broad signification and are merely provisional, and it seems desirable to retain these terms and to employ them in cases where doubt exists, rather than run the risk of encumbering petrology with descriptions apparently precise, really worthless. Unquestionably the microscope is a most valuable assistant when judiciously used, and its application to this branch of natural science promises to give a rapidly increasing knowledge of the mineral composition and minute structure of eruptive rocks.

Many interesting facts connected with the paragenesis of minerals are disclosed by this method of research, and as our knowledge becomes more extended we should look upon these points not merely as questions of purely scientific interest, but we should carefully amass and group our facts in the hope that hereafter they may elucidate some general laws in nature, of which we are not yet cognizant. It would for example be a work of highest interest to map the eruptive rocks and those which have undergone well-marked metamorphism on a large scale, and to note their most important variations in mineral composition. Subsequently the metalliferous lodes might be accurately laid down, and their characters compared with those of the adjacent rock through which they pass. A map carefully constructed upon such a plan would possess considerable scientific interest and might prove to be of great value in mining, since the strongholds of many of the metalliferous lodes are the eruptive and metamorphosed rocks, and if in time to come it should be found possible to establish a definite classification of the association of certain metalliferous lodes with rocks of a particular mineral composition, the microscope will have done as good work for the cause of mining enterprise as for that of scientific truth.

It is well also to bear in mind the fact that the mineralogical composition, the state of aggregation of the components, and other physical characters of a rock, serve to indicate its utility or worthlessness as a building stone; and although it may be argued that nature already demonstrates this in the weathered surfaces of rocks, still we must remember that the agencies most destructive to building stones in large towns do not at all fairly represent the purely natural work of decay which takes place round about the quarry. A good instance of this is afforded by the mineral apatite which is of common occurrence in many eruptive rocks. In these, when naturally weathered, the apatite scarcely seems to undergo any change, yet we know that if acids were present in any quantity in the atmosphere (as is always the case in large towns), it would suffer considerably and in time become decomposed.

So far as the origin of rocks is concerned, the microscopic examination of their sections tends at times to throw some light on questions of this nature. The deductions of Sorby, Rénard, and others upon the temperature and pressure under which certain rocks have solidified, are the result of calculations based upon the relative size of bubbles and crystals contained in the fluids which occupy minute cavities in quartz, and other mineral components of certain rocks.

With regard to the numerous rocks which from time to time have been described as volcanic ashes, tuffs, &c., great difficulty exists in reconciling some of these descriptions with the microscopic appearances which many of them present. In these questions the microscope opens out a large field for scepticism. The fragments of different minerals and rocks are of course to be looked for in deposits of this class. The angular or rounded surfaces of such fragments do not, however, afford us much help in coming to any satisfactory conclusions, since, when the surfaces of the fragments are rounded, we may attribute that rounding to attrition against other fragments, either in the throat of a crater, or to the attrition which water-worn fragments have undergone. Again, the angularity of the fragments will not always serve as evidence of the history of the formation of the rock which is composed of, or which contains them, since we meet with angular fragments in unquestionable volcanic ejectamenta, and also in rocks whose origin has clearly been an aqueous one. In the latter case it is true that the component fragments of the rock are usually water-worn and rounded; but in instances where the materials composing the rock have only been transported a short distance and deposited near the source from which they were derived the rounding consequent upon severe and protracted attrition is not to be expected.

Again, the fact, that the fragments are those of minerals which compose eruptive rocks is no criterion that the rock containing them has also had an eruptive origin. The felspars which occur in the millstone grits and in other sedimentary rocks afford good examples of this, and it needs no proof to show that in many such cases the development of these crystals has not been due to any metamorphic action; moreover the felspars and other minerals of eruptive rocks when occurring in sedimentary deposits are usually in a fragmentary condition, and this would not be the case if they had been developed in situ.

In some instances volcanoes eject showers of fine dust, in which scarcely any trace of definite minerals can be discerned, and the mud thrown out from certain craters appears also to be devoid of fragments or granules which can be referred to any of the minerals which constitute crystalline eruptive rocks. Here again a difficulty arises, for we have apparently no means of discriminating between these unquestionably volcanic products and ordinary sedimentary matter. Lamination again is an equally useless character, for we may have volcanic ashes deposited as sediment in seas or lakes, and intermixed with or containing intercalated strata of ordinary sediment, brought into those seas or lakes by rivers.

These are facts well worthy of consideration, and although in this Memoir I have had occasion to speak of certain rocks as volcanic ash, and although evidence in many cases seems to favour such a supposition, I have done so partly from deference to the opinions of Sir Henry De la Beche, partly from a want of evidence to *disprove* that they are composed of volcanic ejecta-menta, and partly from my own belief, or bias. At the same time I am also prepared to believe that these "ash beds" are often in great part composed of ordinary sediment, while a similar opinion is indicated in Sir Henry's own writings. Before quitting the subject of clastic rocks, it may perhaps be well to point out what may occasionally prove a source of difficulty to the microscopic observer. In some instances, and notably in parts of sections of some of the Brent Tor rocks, it appears that patches of rock, different in character from that surrounding them, shade off gradually into the adjacent matter of which the section is composed, or, in the case of very finely vesicular fragments, they seem to break up into little flecks and granules of irregular form, suggestive of the partial disintegration of these lapilli. This in many cases would I believe be an erroneous interpretation of the phenomenon, which is probably due to the planes of section including only portions of the outer crusts or surfaces of irregularly shaped fragments, the separate patches becoming so thin at their margins that they seem to shade off gradually into the adjacent rock, while, when the lapilli are vesicular and spongy, actual separation of the outlying intervesi-

cular matter occurs.   The accompanying diagrams, Figs. 2 and 3,
will, I think, sufficiently explain what is meant, the parallel lines
in the upper figures representing the planes of section and the
diagonally shaded portions indicating the possible forms of the
imbedded fragments before the sections are cut.

Fig. 2.

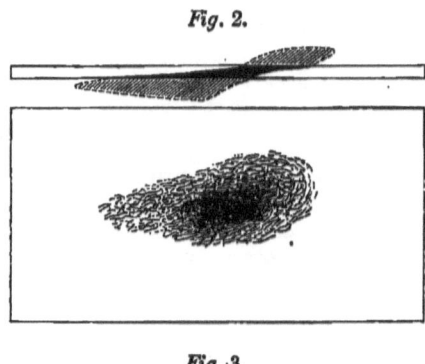

Fig. 3.

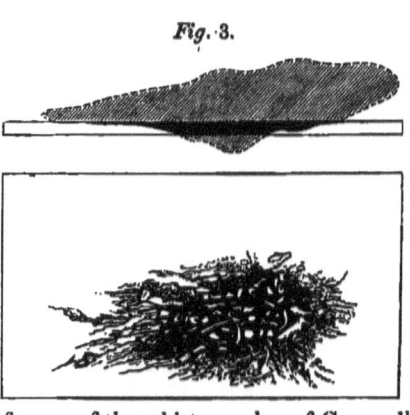

　　Speaking of some of the schistose ashes of Cornwall, Sir Henry
De la Beche observes : " They consist of schistose beds strongly
reminding us of the substance of greenstone, finely comminuted
and permitted to settle in water, in which calcareous matter
was occasionally present.   The principal mineral substance has,
as Dr. Boase observes (Contributions towards a Knowledge of
the Geology of Cornwall; Trans. Geol. Soc. of Cornwall, Vol. IV.,
p. 175), an external character between hornblende and chlorite.
These rocks form part of a general series of variable appearance,
being sometimes slightly coherent, reminding us of the volcanic
ash of modern times ; at others vesicular, or even graduating into
greenstone and other trappean rocks, some of which seem to have
been in a condition of pumice, prior to the infiltration of carbonate
of lime and other substances into the cells which has now rendered
the whole a solid mass."—Rep. Geol. Cornwall, Devon, and W.
Somerset, p. 57.

And again : " There is in fact so intimate a mixture of compact and schistose trappean rocks with the argillaceous slates, that the whole may be regarded as one system, the two kinds of trappean rocks having probably been erupted, one in the state of igneous fusion, and the other in that of ash during the time that the mud, now forming slates, was deposited, the mixtures being irregular from the irregular action of the respective causes which produced them ; so that though the one may have been derived from igneous action, and the other from the ordinary abrasion of pre-existing solid rocks, they were geologically contemporaneous."—Rep. Geol. Cornwall, Devon, and W. Somerset, p. 57.

I have, in this introduction, only dwelt briefly upon the microscopic characters of a very few of the most common minerals which help to form eruptive rocks, and of a few of the most interesting questions relating to those rocks themselves. To give anything beyond this very bare sketch of the characters of the rock-forming minerals of frequent occurrence, would entail the conversion of this introduction into a small text-book of micro-petrology, while it would add greatly to the bulk of the Memoir, and prove most wearisome reading to anyone not specially interested in this branch of study. I have endeavoured very briefly to point out how far the evidence to be derived from a microscopic examination of rocks can be relied on, and how far the older methods of determination formerly, and now, at the field geologist's disposal, can assist and in some cases rival the work performed by more elaborate appliances.

The microscopic examination of rocks not unfrequently subverts preconceived notions regarding their mineral composition, in some cases indicating differences where, to unassisted vision, no differences are apparent, in others showing that rocks very dissimilar in aspect have an identical or similar mineral composition. As instances of the latter I may cite the quartz porphyry, or granitic dyke occurring in Shilla Mill Quarry near Tavistock, the rock although varying considerably in colour and structure in different parts of the quarry being identical so far as the component minerals are concerned. Again, the dark iron-grey and patently micacaeous dykes of the Lake District are composed of the same minerals as the brick-red dykes which are so common in the Upper Silurian rocks of that country. In the latter, however, the magnesian mica is but poorly represented, so that while the one rock would be described as minette or mica-trap, the other would almost be designated micaceous felstone.

Rocks intermediate in character between basalt and trachyte occur in several parts of the Lake District and Wales, associated in both instances with sedimentary deposits of silurian age. The microscopic characters of those pertaining to the former localities have been described by my colleague Mr. J. Clifton Ward, while those of the latter I have only recently investigated by order of Professor Ramsay. Mr. Ward's conclusions and my own coincide very closely, and indicate a community in the mineral composition of these approximately coeval but now widely separated lavas.

Various forms of goniometers have from time to time been devised for measuring the angles of crystals when seen under the microscope. Of these that perhaps known as Schmidt's goniometer is the simplest in construction ; but, when the section in which the crystals occur is rather thick or dark, it is often difficult to see the cobweb with sufficient distinctness to enable the observer to adjust the instrument with precision. It is, moreover, well to remember that we can seldom be sure that the sections of crystals imbedded in thin slices of rock are precisely at right angles to the faces whose included angle it is requisite to measure ; so that even at the best we cannot obtain more than approximate measurements as a rule, and these measurements are only available for purposes of distinguishing between minerals when their respective angles vary to a considerable extent. To effect crude measurements of this kind it often suffices to draw two lines on paper, by means of the camera, corresponding with the outlines of the faces seen in the section, and then to measure off the angle formed by the intersection of these lines with a protractor. This simple method, however, will only give results to within 30 minutes, or at best 15 minutes, when a protractor of ordinary size is used ; and even if a more perfectly graduated instrument were employed it is doubtful whether better results would be obtained, owing to the difficulty experienced in drawing the two lines with any great amount of precision, as the boundaries of the crystals are often rough and irregular. Indeed when this method is adopted it is usually desirable to draw the same lines several times until they give similar readings with the protractor. Where very precise measurements are requisite the observations should be made with an ordinary reflecting goniometer upon small isolated crystals, i.e., if it be possible to isolate them from the rock in which they occur.

In some cases it is desirable to etch the smoothly-ground surfaces of chips of rock with acids. By this means information may occasionally be acquired, not merely with regard to the solubility of certain mineral components, but at times evidence may be procured of a somewhat different nature. For example, a rock from Wrington Warren, near Bristol,* was found, after undergoing such treatment, to contain fragments of limestone in which portions of minute crinoid stems were imbedded ; the limestone fragments forming little angular patches in a basalt.

The application of acids to thin sections under the microscope is rather a difficult process, especially when the rock contains carbonates, and it is needful to examine the deportment of other adjacent minerals with acids. The difficulty in such cases consists in restricting the drop of acid to any particular crystal or fragment. Of course in the case of mounted sections covered with glass it is necessary to remove the cover and clean off the superjacent balsam before any reagent can be applied, and this

---

* Described in the Appendix to the Geological Survey Memoir on the Bristol and Somersetshire Coalfields 1876, p. 208.

frequently entails the destruction of the section. The best way is to apply the acid before the section is reduced to its ultimate degree of thinness; the acid can then be washed off, the process of grinding continued, and the section finished in the usual way; and even in this case there is risk of affecting a greater thickness of the section than might be anticipated.

A few words may here be said on the preparation of microscopic drawings. In all cases it is desirable to make the outline by means either of a neutral-tint glass reflector, or by an ordinary camera (Wollaston's prism). The former entails either a reversal of the object or a transfer of the outline, the latter involving additional labour, while the former necessitates the filling in of the detail of an object seen through a thick glass slide. The Wollaston's prism is therefore best adapted for drawings of rock sections. Some observers dispense with the use of any instrument of this kind and make the outline by eye, placing the paper on a level with the stage of the microscope. The detail should always be filled in by eye alone, a moveable index placed in the eye-piece of the microscope serving to record the precise spot where the work is being carried on. Some draughtsmen are in the habit of rendering their drawings diagrammatic, but this practice, although it may have its advantages in forcibly depicting the views of the observer, detracts from the value of the drawing as a truthful delineation of the object. In this Memoir I have tried to make faithful copies of the rocks as seen under the microscope, so that, should my conclusions be doubted, my drawings may at all events be beyond question.

In most cases the sections examined and described in this Memoir have been prepared by Mr. Cuttell, of New Compton Street, Soho; a few I have cut for myself. Machines constructed for this purpose may now be seen at the Loan Exhibition of Scientific Apparatus at South Kensington, while descriptions of how to prepare sections of rocks and minerals will be found in the following works: "How to Work with the Microscope," by Dr. Lionel Beale; "The Microscope in Geology" (an article in the Popular Science Review, No. 25, Oct., 1867, p. 355), by the late David Forbes; and an article by J. B. Jordan, in the Journal of the Quekett Microscopical Club.

It is greatly to be regretted that at present our literature upon micro-petrology is very limited, and for this reason it is most desirable to examine eruptive rocks from foreign localities, to compare them with the descriptions given of them by different continental petrologists, and to ascertain as far as possible in what respects they may be correlated lithologically or otherwise with the rocks which occur in our own country.

In investigations of this nature it does not suffice to rely implicitly upon microscopic research alone. The labours of the field geologist, the crystallographer, the chemist, and the physicist should all be impressed into the service, and it is for this reason that the work of an individual observer is not likely to be based upon a thorough knowledge of all these different branches of science; while unless

B

the sum of such special investigations can be added up, and the deductions which are based upon them duly weighed, we can only regard our petrological conclusions as provisional, although in many cases they may ultimately prove to be close approximations to the truth. Petrology is but a branch of geology, yet petrology is itself a composite science, while to *thoroughly* follow it out in all its ramifications is a labour which of necessity exceeds the capabilities of any single observer. It represents a large field for inquiry. Its great aim an elucidation of the origin of rocks by careful study of their mode of occurrence, their mineral composition, and their structural characters, both gross and minute. It is in fact the inanimate history of the globe based upon the anatomy and histology of the earth's crust. Apart from its scientific interest it has practical bearings of great industrial importance, but the scientific knowledge must be accumulated and arranged before it can be applied successfully to the interests of the community at large.

The following is a rough explanatory table of the mineral components, &c. of the rocks mentioned in this work :—

BASALT (Dolerite, Anamesite, Melaphyre).	Plagioclase. Augite. Magnetite (Titaniferous iron), Apatite, Olivine.
DIABASE - -	- Plagioclase (Oligoclase according to Dathe). Augite, generally represented by pseudomorphs of serpentine, &c. Magnetite, Titaniferous iron, Apatite, Olivine. Chlorite.
DIORITE - -	- Plagioclase. Hornblende. Magnetite (Titaniferous iron), Apatite, Quartz.
GABBRO - -	- Plagioclase. Diallage. Magnetite (Titaniferous iron).
GREENSTONE	- A term used of late years as a synonym for Diorite, but formerly embracing all varieties of Basalt, Diorite, and Gabbro. In this work it is used in the earlier sense of the term.
APHANITE -	- Any fine-grained varieties of the preceding rocks which, owing to the imperfect or minute development of the individual mineral components, cannot be determined with precision.
HYPERSTHENE ROCK	- The rocks formerly described under this name are now, for the most part, regarded as Gabbros, the Diallage or Enstatite which they contain having been mistaken for Hypersthene.
AMPHIBOLITE (Hornblende Rock.)	Hornblende. Quartz.

GRANITE - 　　　- 　- Orthoclase (Plagioclase).
　　　　　　　　　　　　Mica (Potash or Magnesian, or both).
　　　　　　　　　　　　Quartz.

GRANULITE 　　　- 　- Orthoclase (Plagioclase).
　　　　　　　　　　　　Quartz.

QUARTZ 　　PORPHYRY 　Orthoclase.
(Elvan).　　　　　　　　Micas.
　　　　　　　　　　　　Quartz.
　　　　　　　　　　　　Felsitic magma.

PITCHSTONE (Rhyolite in 　Vitreous rock, variable in composition, but
part).　　　　　　　　approximating to felspars and usually con-
　　　　　　　　　　　　taining water, the vitreous condition of rocks
　　　　　　　　　　　　of this class being due to rapid solidification.
　　　　　　　　　　　　The representatives of these rocks at Brent
　　　　　　　　　　　　Tor have undergone considerable physical if
　　　　　　　　　　　　not chemical change, and are now in great
　　　　　　　　　　　　part devitrified by the development of
　　　　　　　　　　　　Microliths.

RHYOLITE 　　　- 　- A term which embraces Quartz-trachyte, Ob-
　　　　　　　　　　　　sidian, Pitchstone, &c. The Rhyolites are
　　　　　　　　　　　　classed as granitic, felsitic and hyaline. In
　　　　　　　　　　　　this work rocks pertaining to the two last-
　　　　　　　　　　　　mentioned classes are alone implied.

CLASTIC ROCKS - 　　- These include volcanic ashes and sedimentary
　　　　　　　　　　　　deposits composed wholly or in part of frag-
　　　　　　　　　　　　ments of rocks and minerals.

SCHALSTEIN 　　　- 　- A rock which is possibly related to Diabase in
　　　　　　　　　　　　composition, has a schistose structure, and is
　　　　　　　　　　　　frequently amygdaloidal.

ALTERED GREENSTONE - This expression is used to denote eruptive rocks
　　　　　　　　　　　　so far decomposed that their normal mineral
　　　　　　　　　　　　constituents can only be partially recognised.
　　　　　　　　　　　　They were probably once Basalts or Diorites.

PART I.

EXAMINATION OF THE ROCKS IN THE FIELD.

The area embraced in this Memoir has for its boundaries the River Tamar on the west, and the western margin of the Dartmoor granite on the east, while lines drawn east and west, about a mile south of Lidford and a mile north of Calstock, represent the northern and southern limits.

These purely arbitrary boundaries comprise about 60 square miles, the area including only portion of a much larger district in which beds of volcanic ash and bosses and dykes of other eruptive rocks occur in great number, and often represent continuations of the same beds, or spurs from the same deep-seated intrusive masses.

Within the small district of which this Memoir treats, there occur many highly interesting and peculiar eruptive rocks, some intruded through, and others interbedded with, the Carboniferous and Devonian strata which lie on the west of Dartmoor.

The object now in view is an elucidation, to some extent, of the mineral composition and structure of these rocks, not merely by examination of them in the field, but also by more minute investigation of carefully selected specimens ; thus supplementing the work of the late Sir Henry De la Beche by methods of research unknown to petrologists in the days when he surveyed this district, and rendered it classic ground by the careful observations which he made and published in his Report on the Geology of Cornwall, Devon, and West Somerset. The physical features of this district need but a brief description here, since they are admirably dealt with in the work just alluded to.

On reference to the Geological Survey Map, Sheet XXV. (1-inch scale), it will be seen that the culm or carboniferous series and the Devonian series of rocks abut abruptly against the western flanks of the granite which constitutes Dartmoor, and that from the line of contact a light, graduated wash of the colour, used to denote granite, is carried for a little distance over the adjacent culm and Devonian rocks. From this, however, it should not be inferred that these sedimentary deposits approximate either in mineral composition or in lithological character to the granite itself; in other words, it must not be supposed that there is a gradual passage from these rocks into granite, since there is little evidence, so far as I have seen, at all events near the present denuded surface on the east of Brent Tor, to show that such is the case. In the absence of opportunity for more detailed investigation it is, however, only right to quote the following statement which occurs in Sir Henry De la Beche's Report, p. 267 " In numerous localities we find the coarser slates converted into rocks resembling mica-slate and gneiss, a fact particularly well exhibited in the neighbourhood of Meavy, on the south-east of Tavistock." Where exposures of culm measures or Devonian

Fig. 4.—Ideal Section from the W. Edge of Dartmoor to Ridge.

rocks occur near the margin of the granite, appearances usually indicate a slight physical change rather than a mineralogical differentiation. Approximations to this latter phase are, however, perceptible in places in the form of imperfectly developed crystals, which may in some cases represent staurolite. About half a mile to the S.W. of Cock's Tor the grits have undergone some alteration and hornblendic crystals have been developed in them. It happens, unfortunately, that where lines of actual contact exist, they are generally masked by superficial accumulations; but it is often possible to procure samples of the rocks within a few yards of the contact, and if they then exhibit no alteration, it is evident that it is either of very trifling extent or that the demarcation between the two rocks is a comparatively sharp one.

On referring to the Geological Survey Map it will be seen that at two localities, Brazen Tor and Waspworthy, there are small patches, mapped in as greenstone, which are in actual contact with the granite. These are Hornblende Rock (Amphibolite), and the line of demarcation between it and the granite is a very sharp one, and corresponds exactly with the line drawn on the map at the former locality; but at the latter the evidence is not so clear, as there is very little rock exposed. A little further to the south-west two large intrusive masses of somewhat similar character constitute the features known as Smear Ridge and White Tor. These, together with the Brazen Tor and Waspworthy patches, are probably connected at some little depth beneath the present surface. The ideal section Fig. 4. may serve to illustrate this. Cock's Tor situated a little further to the south is composed of a rock very similar in appearance to those just mentioned, but under the microscope it is seen to contain little or no free silica, and to be composed of diallage, plagioclase, titanic iron, and apparently a little apatite. It must therefore be regarded as a Gabbro. In support of this view, I may add that Professor A. Rénard of Louvain has also examined this rock, and states that many unquestionable Gabbros which occur in Belgium, resemble it in every respect.* This rock extends due west on the map, but is much obscured by superficial accumulations on the west of the tor.

* Height of Cock's Tor, 1,472 feet; Brent Tor, cir. 1,100 feet; Black Down, 1,160 feet above the sea level.

At Sowtentown it is represented as bifurcating and spreading westward in two broad and long strips. I am inclined to think that it does certainly cross the valley of the Tavy, but some parts of the southern prolongation yield rocks of a totally different character, and although Sir Henry de la Beche, when describing them in his Report,* points out many of their changes; he appears to have entertained some doubt about the exact relation which these rocks bear to one another, although he has included them within the same boundary lines on the Survey Map. To ascertain their precise relations to one another, so as to map them separately, would certainly be a somewhat difficult task, and the work would have to be executed upon a larger scale than one inch to the mile.

At the Cottage Inn, on the main road north-east of Tavistock, there is a good exposure of Gabbro, succeeded further along the road by black gritty shale much veined and containing little cubes of pyrites, but the contact is not visible. Beyond this, close to a bend in the road, what appears to be a slaty breccia occurs, in contact with a schistose rock which resembles some of the beds described as ash. I believe that Gabbro, similar to that occurring at the Cottage Inn, extends as far as Indescombe, where another exposure is visible. Near this point the belt mapped as greenstone is constricted, while further on it becomes schistose, and may in many cases represent consolidated ash beds. I am even disposed to think that the constriction shown on the map may amount to an actual separation of these rocks, which apparently have nothing in common to justify their being mapped in the same patch, the one being intrusive, while the other is interbedded and schistose in character, nor can I conceive a passage between rocks so dissimilar. That they may inosculate seems possible as the two strips trend in the same east and west direction, and it may have been from some such inference that they were originally mapped together. The difficulty of ascertaining the boundaries is considerable, the exposures not being very numerous.

There are, however, one or two things to be said in defence of this piece of mapping :—

First. That the schistose rocks were called greenstone ash and were mapped with the greenstones, while the broad signification which the term greenstone possessed some years ago, embracing as it did diorites, basalts, &c., serves to disarm criticism.

Second. The small scale of the map rendered the insertion of much detail inadvisable.

Third. The details relating to lithological differences are given in the "Report on Cornwall, Devon, and West Somerset,"

---

* "Where greenstones approximate or come into contact with the granite, the "crystallization has become large near it, and even a modification of the arrange- "ment of the component particles seems to have been produced, so that the horn- "blende resembles hypersthene."—Geological Report on Cornwall, Devon, and West Somerset, p. 268.

although such information is not embodied in the map on account of the small scale upon which it is constructed.

Some of Sir Henry De la Beche's remarks upon these rocks will be cited further on, together with a few additional observations.

The northern branch of the Cock's Tor fork extends westward for a distance of over seven miles, and although in some parts the exposures are few and unsatisfactory, still the evidence tends to show that the belt is of a more or less schistose and ashy character throughout a considerable portion of its length, and it therefore becomes a question whether *this* strip should be represented as actually joining the spur of gabbro which passes westward from Cock's Tor. That the two strips proceeding from this spur are denuded portions of the same beds I have little doubt, as in both cases they are in great part composed of schistose ash with some local variations in structure, texture, and mineral composition, while from the paucity of exposures in some places it is scarcely possible to note the gradual passages which may take place between them. North of these strips outliers of very similar character occur, and on reference to the Geological Survey map it will be noticed that the principal ones assume a somewhat annular disposition around the mass which constitutes Brent Tor.

· The strip which lies immediately south of Brent Tor, and which passes west from Burnford Farm through Pillands and Churlhanger, is shown on the map to branch in a northerly direction a little north of Foghanger, and about a mile further north it bends again into a western course extending slightly beyond the Tamar and ending abruptly a little south of Greston Bridge. From the dips recorded on the Survey map, and from one or two additional ones, it seems probable that the three strips of ash which cross the main road from Tavistock to Brent Tor are but repetitions of the same beds, while their tolerably persistent character in no way tends to invalidate such a supposition. The great doubt expressed by Sir Henry De la Beche about the boundary between the culm and the Devonian series also serves to render this question an open one. Towards Dunterton the ash seems to give place to a compact rock resembling basalt, and at Greston Bridge the isolated patch marked on the map is shown to consist of two beds of a similar rock, parted by and resting upon black shales approximating to Lydian stone, Pl. VI. fig. 4.

The strips of ash on the north and east of Brent Tor, although showing local variations, may also be regarded, with a fair amount of probability, as being continuations of the beds on the south and west. Brent Tor itself differs considerably from these ash beds which encircle it. On ascending the Tor one is struck with the compactness of some of the rocks and with the slag-like vesicular appearance of others, as compared with the fissile schistose character of the neighbouring ashes. Imbedded fragments and lapilli, frequently rounded, project from the weathered surfaces, and although the rocks which compose the Tor appear at first sight to be slightly varying conditions of the same kind of rock, yet they form as it were a group by themselves, and may be clearly distinguished from the ash beds which environ them. We shall subsequently enter into some considerations connected with the granite of Dartmoor and Hingston Down and the Elvan dykes, which proceed from the granite masses, but at present it will be better to restrict our attention to Brent Tor and the ash beds which surround it.

Brent Tor is a small hill with a rather gradual slope upon the eastern, and more abrupt ones on the northern and southern sides, while on its western or north-western aspect it is craggy and precipitous. (Plate VI. page 31, figs. 2 and 3.) It is a conspicuous object for many miles around, but this is due to its occupying an elevated situation, the ground from which the tor arises sloping gradually, but almost constantly, downwards in the direction of Tavistock, while on the three remaining sides it also declines, slightly on the north, but on the east and west into valleys of considerable depth. Viewed from a distance, as from some spots on the Morwell Downs, it appears to stand as an isolated hill upon a plateau (Plate II.), and I am disposed to think that this is the case, and that Brent Tor is a hard mass which has offered considerable resistance to degrading agents, and that it is surrounded by on old plane of marine denudation. Viewed from the south-west a decided dip is apparent in the beds which constitute the tor, as shown in Plate III., but the dip varies, appearing to curve upwards towards the northern side. This dip is more perceptible at some distance from the tor than when the observer is in its immediate vicinity.

At the base of the tor, on the N.N.W., an aphanitic-looking rock occurs, which is probably the "greenstone" mentioned by Sir Henry De la Beche (Report on Devon and Cornwall, p. 121). From microscopic examination I am inclined to regard this rock as a much decomposed basalt, with an originally glassy magma which is now devitrified. On ascending the tor this is succeeded by another rock very similar in appearance, from the weathered surface of which angular and sub-angular fragments project. The latter are mostly fine-grained and resemble the matrix, which is here and there vesicular. Still higher on the north side the rock seems persistently fine-grained and aphanitic in appearance, the weathered surfaces showing in places a brecciated structure, but with the fragments resembling the matrix. At one spot at the base of the tor, on the north-western side, small slaty-looking patches are imbedded in the fine-grained rock. About half way down the south side of the tor, one or two of the beds appear to be scoriaceous on the weathered surfaces. When this weathering has more completely decomposed the contents of the vesicles the rock resembles a cinder, but where less weathered the fresh fracture merely shows a light spotted aphanitic-looking rock, somewhat similar in matrix to those already observed.

In one exposure of rock on the lower part of the south side of the tor, strong vertical joints are visible running about north and south, and 15° south of west.

PLATE II.

BRENT TOR FROM HART'S HALL, NEAR MORWELLHAM.

PLATE III.

BRENT TOR.—VIEW TAKEN FROM ABOUT HALF A MILE TO THE SOUTH-WEST.

In the field on the south side of the tor slightly schistose rock occurs, with projecting nodules on the weathered surface. These nodules are sometimes fine-grained and compact like the matrix, but they are usually scoriaceous, the vesicles often being filled with a white zeolitic mineral. The sandy-looking nodules frequently contain irregular scoriaceous patches, and lower down in the field a rock occurs with included fragments resembling slate and red jasper. The appearance, however, is deceptive, for by microscopic examination the rock is seen to be a partially devitrified pitchstone, while the imbedded fragments are finely vesicular lapilli resembling pumice. Still further down the hill, close to South Brentor village, there are blocks which contain slaty-looking fragments and scoria.

Some of the rocks which constitute the tor present, under the microscope, a decidedly rhyolitic aspect and approximate in character either to the felsitic or the hyaline rhyolites. They are for the most part micro-felsitic; but it is possible that they were originally hyaline and that their felsitic character is due to devitrification.

Upon first examining Brent Tor it appeared to me probable that the rocks of which it consists might merely represent part of an old bed or series of beds of sediment into which scoria and lapilli had been showered, but further inspection, coupled with a microscopic examination of the rocks, seems conclusively to show that the statement of Sir Henry De la Beche was a perfectly correct one, viz., " That in the vicinity of Brentor a volcano " had been in action, producing effects similar to those produced " by active volcanoes from a similarity of causes, forcibly presents " itself."[*] Had there been a wide spread of rock like that which forms Brent Tor, it would from its thickness and superior hardness have offered great resistance to denuding agents in other parts of the neighbourhood, for as there is no violent contortion or folding of the beds in this area, but simply gentle curvature, it seems highly improbable that, in the absence of faulting, Brent Tor merely represents the last trace of a wide-spread deposit. Conclusions upon this subject might be more easily arrived at if exposures of rock in the vicinity were more numerous and dips could be determined with greater precision; but unfortunately the lie of the rocks is very obscure in some places.

Having thus far described what is to be seen upon the ground at Brent Tor itself, it is now advisable to give some account of field observations made upon the various beds of volcanic origin which occur around, and which possibly emanated from it. These beds are mainly schistose in structure; but whether this is to be attributed to original deposition, or whether it may in part be due to the pressure of an enormous mass of superincumbent rock which once undoubtedly overlaid them, appears to me to be an open question. In other words, is this schistose structure one which has been superinduced by vertical pressure, and actually a

---

* Rep. Geol. Cornw., Dev., and W. Somerset, p. 122.

cleavage which is more or less coincident with the bedding; cleavage transverse to the bedding planes being absent simply because the rocks have, so far as dips can be noted, undergone but little pressure laterally? The persistence of this structure over a considerable area would fit in equally well with either hypothesis; but the almost equally persistent vesicular character of these beds, and their strong resemblance in many instances to doleritic rocks, seems to lend some plausibility to the assumption that the schistose structure may be due to a kind of cleavage. That these beds are in places of an ashy character seems probable, since fragments of felspars, lapilli, and scoria are frequently met with in them, although they are often only sparsely distributed. We must, however, always bear in mind the possibility of such beds resulting partially or wholly from the degradation of pre-existing eruptive rocks.

In the immediate vicinity of Tavistock, and to the east of the town, two strips of schistose rock occur which appear in no way to differ from much of the ash which lies between Tavistock and Brent Tor, and it is quite possible that these two strips are merely repetitions of the same beds. The strip which stretches up to Mount Tavy, on the eastern side of the river, dips in a N.W. direction; the dip of the other strip, which occurs nearer the town, is not so evident. On the road leading out of Tavistock in a westerly direction, just before reaching Downhouse Farm, killas is exposed in the roadside; it is somewhat contorted and affords no trustworthy dip. At the cross roads about a quarter of a mile north of Stiles Wick a rock occurs containing white flecks of kaolinised felspar and hornblende : it is apparently a decomposing diorite. The exposures of rock about this spot are scarcely worth consideration, and I am not sure that they represent rock in place. At Downhouse farm I picked up a piece of vesicular ash in the road, but could not see any in place. In the side of a lane leading into the main road exposures of scoriaceous ash and killas may be seen, but it is very doubtful whether they are in place. Probably they are merely loose pieces imbedded in the bank. The ash contains cubes of decomposing pyrites, some of them half an inch in diameter. In the main road from Tavistock to Milton Abbot there is a good exposure of schistose ash, which at Langford Farm shows interbedded vesicular bands (Fig. 5) about three to

Fig. 5.

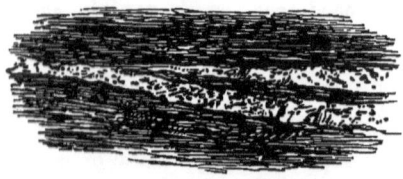

four inches in thickness, lying in a somewhat horizontal position, but the beds appear to dip to the S.S.W. at a low angle. I am inclined to think that the Hardwick and Langford Farm strip of ash terminates at the constriction shown on the map, in the neighbourhood of Indescombe.

We will next consider the belt of rock which runs in an east and west direction by Ridge, Kilworthy, Lamerton, and Shortaburn. At Ridge, on the road from Tavistock to Petertavy, it seems closely to resemble that exposed at the Cottage Inn on the main road on the western bank of the Tavy. It therefore approximates more nearly to a gabbro than to an ash, and consequently I think that there ought to be a separation of this rock from that which occurs further to the west, although a continuous band is coloured on the Survey map. It is possible that a passage may occur between these crystalline rocks and the schistose beds, but it is more probable that this is not the case. Near Kilworthy, *i.e.*, between Hardwick and Kilworthy, there is a fine quarry on the east side of the road leading from Tavistock to Brent Tor. It is a vesicular ash, manganese oxides occurring in the vesicles and in small druses and cavities. Balls of a harder and more compact stone are met with in these beds. According to one of the quarrymen these balls lie in definite layers, but, although I saw several in place, I could not find evidence enough to corroborate this statement. By the same authority I was told that a band of hone-stone occurred to the south of the quarry, but I was unable to find any exposure of it. The quarry shows no killas, either at top or bottom, and represents a thickness of over 20 feet of ash. It is being worked as a building stone, and is almost exclusively used for the construction of Kelly College. A specimen picked up in the quarry shows rounded vesicles as large as a walnut. The rock seems to vary but little in lithological character throughout the quarry, although in some places it appears more like a decomposing vesicular basalt than an ash. It is, however, more or less schistose in character throughout. It is traversed by numerous joints and is a soft stone to work. By the roadside just opposite Lamerton Church, there is a small exposure of rubbly, decomposed ash, which when much decomposed becomes a yellowish, ochreous earth. About Shortaburn there are no exposures of ash visible, but some of the loose stones and road metal in the neighbourhood contain scoriaceous fragments. Near Twowell the loose stones built into the banks and hedges, and lying in the roads, are vesicular ash much decomposed, the cavities containing manganese oxides. No exposures of rock in place are visible either here or in the fields about Shortaburn, but there is a tolerably good feature in places which no doubt served to determine the boundary lines drawn on the map. Some exposures of ash are mentioned by Sir Henry De la Beche as occurring near Combe.

In the belt of ash lying north of that last mentioned, the best outcrops appear to be in the neighbourhood of Churlhanger. In the roadside, just S.W. of the farm house, there is an exposure of five or six feet of a vesicular schistose rock: the latter character gives a peculiarly jagged appearance to the exposed surface.

Fig. 6. It is full of small specks and granules of calcspar, and where these have been weathered out the rock is seen to be

Fig. 6.

finely vesicular. In places larger amygdaloids of calcspar occur, and also narrow calcspar veins. The schistose structure appears to coincide with the bedding, which in one place is nearly horizontal. In a field close to the road, but at a considerably higher level, there is also a good exposure of the rock. It is much the same in character as that in the road beneath. Nodular-looking patches weather out in relief on the surface, and show a finely spongy, vesicular structure, which upon fracture is seen to extend half an inch or more from the surface, but further inwards the vesicles are filled with calcspar. This rock appears to be identical with some of the schalstein which occurs in Nassau.

ELVAN DYKES.

To the south and south-east of Tavistock several elvan dykes are marked on the map. I was only able to visit the two northernmost, which are mapped as trending due east and west. On attempting to follow out the northern one from Morwell Rocks, I was unable to find any exposure of the dyke until reaching Shilla Mill, where there is a good quarry, now being worked for road metal, not far from the banks of the Tavy, and about three quarters of a mile from Ramsland. There is nothing but elvan exposed in this quarry except a little rubble which lies on the top, and which very likely represents the disintegration of the underlying rock. On the south side of the quarry the rock is of a dark grey colour, with irregular dark greenish-black patches imbedded in the grey, granular, felsitic matrix, but further to the north of the quarry it becomes a quartzose felspar porphyry of the usual elvanitic type. In places the rock contains a little pyrites or mispickel, and also specks of copper pyrites and some chlorite. From Morwellham to Ramsland there seems to be no definite exposure of the elvan, and certainly there is none in any of the paths or roads in the neighbourhood. The house and walls at Roman's Lee were built out of this elvan about nine years ago, and the stones lying in the vicinity resemble the darker variety of rock from the northern elvan at Shilla Mill quarry. ·

The best exposure that I could find of the long southern elvan course was at the Lower Grenofen quarry. At this spot the rock has been worked for kerbstones, &c., the consignments having mainly been to London. The quarry is now abandoned. There is an exposure of over thirty feet of elvan. The felsitic matrix of the rock has a pinkish tinge, but loses colour on exposure. Porphyritically imbedded crystals of orthoclase occur plentifully in this matrix, the other components are seemingly mica, hornblende, and chlorite. The rock is irregularly jointed, but there are some well-developed parallel master joints running several feet apart. Plate VI., fig. 1.

The annexed extract relating to the Morwell Down tunnel was copied for me by Mr. G. Chowen of Whitchurch, to whom I was much indebted for information and assistance when in that neighbourhood.

"The following are the strata which have been passed through, commencing at the north end of the tunnel; the provincial names of the different rocks being given as they are generally used by the Cornish miners.

	Fathoms.
Killas, metalliferous slate - -	311
Elvan, chlorite and quartz -	11
Killas - - ·	23
Grouan, clay-porphyry -	6
Killas - - - -	12
Grouan - - -	26
Killas } ,, with veins of quartz }	436
Elvan - - -	15
Killas - - -	3
Grouan - . -	7
Elvan, quartz granular and crystalline -	12
Killas - - -	408
Entire length of tunnel - - 1,270	

"The directions of all these beds seem to be parallel to each other and to range nearly east and west.

"Two facts have been ascertained by its progress. 1st. Relative to the rocks, that the killas of which the hill is mainly formed is traversed by beds of other rocks, the direction of which is inclined to that of the metalliferous veins, and which have a pretty uniform dip or underlay to the north. 2nd. Relative to the metallic veins or lodes that they traverse all the strata, and that they have a remarkable difference in their dip or underlay on the two sides of the hill; those on the north side dipping to the north and those on the south side to the south.

"The lodes near the centre of the Morwell Down hill are intersected by two cross lodes or cross courses. The killas on each side of one of these occurs in a remarkably altered state as a soft clayey substance, so incoherent as to have rendered the preservation of a passage through it until it was securely arched very difficult. Nineteen metalliferous veins and two cross courses were intersected by the tunnel in the Morwell Down hill."—Paper by J. Taylor, Trans Geol. Soc. Lond., vol. iv. 1817.

MARGINS OF GRANITIC MASSES.

The only spot where I have been able to see any definite alteration of the Devonian slates near their contact with granite is at the Morwell Rocks. Here the slates resemble in places finely speckled fruchtschiefer,* the speckles probably representing an incipient development of chiastolite or staurolite, but no definitely formed crystals are to be seen. These speckled slates lie on the opposite side of the Tamar to that on which the granitic mass of Hingston Down occurs, and are apparently about a quarter of a mile from the contact. It is probable, however, that alteration, of this kind occurs in many other localities in the district, but exposures of contact, or in the immediate neighbourhoods of contact of the sedimentary with the eruptive rocks, are few, and the time placed at my disposal for the prosecution of this work precluded the possibility of persistently following the boundary lines in search of such exposures.

Opposite the Morwell Rocks, upon the other side of the Tamar, there is a good quarry of granite worked by the Gunnis Lake Company. It was formerly called the " Tamar Granite quarry " and has been worked for over 70 years with little intermission. The granite is fine-grained, and the orthoclase in it is mostly white, but there is a course which contains pinkish orthoclase, no sharp demarcation, however, existing between this and the other variety. A very fine-grained rock, with a few large quartz blebs or crystals occurs in one part of the quarry (it is called elvan by the quarrymen). The granite gets out in good-sized blocks, and is used for building purposes in London, Exeter, and other places. It seems very free from pyrites, and is apparently a stone well adapted for architectural purposes.

At Brazen Tor† (pronounced " Brauzen Tor "), which is situated on the western margin of Dartmoor, and about six miles north-east of Tavistock, the granite consists of large orthoclase crystals, biotite and quartz, the crystals of orthoclase often exceeding two inches in diameter. On reference to the published geological map it will be seen that there is a patch of rock coloured as greenstone, which is represented as being in contact with the granite. The junction is not actually visible, but the line may be drawn within a few yards. The granite close to the contact is very fine-grained and felstone-like, loose blocks showing black nests and segregations of schorl. It becomes more and more porphyritic towards the tor itself. Here and there quartz veins an inch or two in thickness are seen to cut through the fine-grained granite. Towards the top of the tor the black schorlaceous segregations form projecting knobs on the weathered surface of the rock. The nearest approach to contact shows a rapid change from gabbro into schorl-spotted elvanitic rock or fine-grained granite, and then passes on into porphyritic granite. The line on the map is very correctly drawn. The mineral components of this rock are coarsely crystallized and loosely aggregated. It would not make even a moderately good

---

* The fruchtschiefer of continental petrologists embraces spotted and speckled shales or mica-schists ; the spots consisting of talcose or micaceous, scaly aggregates, or segregations of dark, opaque granular matter, which is probably graphite or some manganese oxide.

† See also pages 38 and 42.

building stone. At the top of Brazen Tor the granite shows very
regular mural jointing. Fig. 7.

Fig. 7.

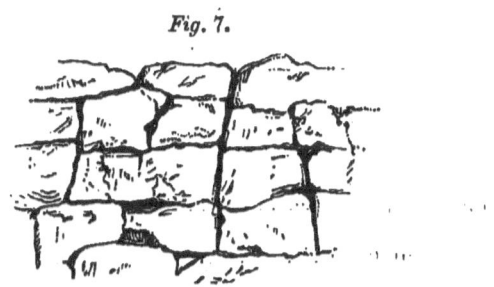

Of the junction between greenstone and granite indicated on
the map, about half a mile to the west of this, I could get no good
evidence. The rock mapped as greenstone occurs close to the
little bridge near Waspworthy, but the actual junction is not
visible. In a field north of the brook, close to Waspworthy, a fine-
grained or compact amphibolite occurs; granite blocks are also
scattered about on the surface. It is possible that the Brazen Tor
patch of amphibolite joins this, but no trustworthy exposures are
to be seen near the road, nor in any of the adjoining fields.

That these patches are also connected at some depth beneath the
surface with the mass lying between Crayston and White Tor and
that which constitutes Smear Ridge, there can be little doubt. In
a field on the east side of the road, close to Horndon, blocks of
gabbro occur in large numbers, some of them of considerable size,
up to 9 feet in height, such as the block locally known as the
"Master Rock," Fig. 8, and some of them apparently in place.

Fig. 8.

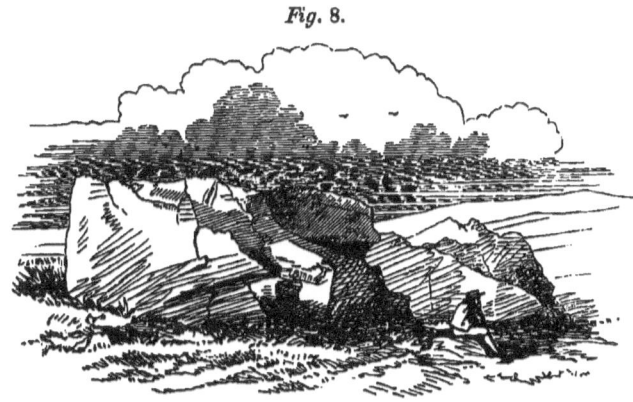

These are outside the boundary of the greenstone marked on the
Survey map, and one would at first sight be inclined to think this
boundary line to be incorrectly drawn; but, on examining an old

grubbing at Kingsford Corner, killas will be found at a depth of
about 12 feet, clearly showing that, if the line can be extended at
all, the alteration will only be a slight one, embracing perhaps a
quarter of a square mile, and that it was drawn carefully, and
within what Sir Henry De la Beche considered safe limits.   In
his Report on the Geology of Cornwall, Devon, and West Somerset,
he refers part of these greenstones to Hypersthene Rock.

The margin of the granite just west of Great Staple Tor consists
of fine-grained rock, which on nearing the tor becomes coarsely
porphyritic with orthoclase crystals twinned on the Carlsbad type.
There is no visible contact exposed between the granite and the
gabbro of Cock's Tor, a moderately deep valley lying between the
two features, Plate IV.   The mica in the granite of Great Staple
Tor is biotite.   The granite composing the Tor exhibits well-
marked mural jointing.   Plate V.

This mural jointing in granite has been noticed by Sir Henry
De la Beche, Dr. Boase, and others, who state that it is especially
characteristic of the large granitic masses where they come in
contact with other rocks.   The former writer observes : " This
variety of cleavage may be regarded as a kind of thick lami-
nated structure pervading the masses on the large scale, probably
agreeing in form with that of their original surfaces after pro-
trusion."—Rep. Geol. Cornw., Dev., and W. Som., p. 163.

PLATE IV.

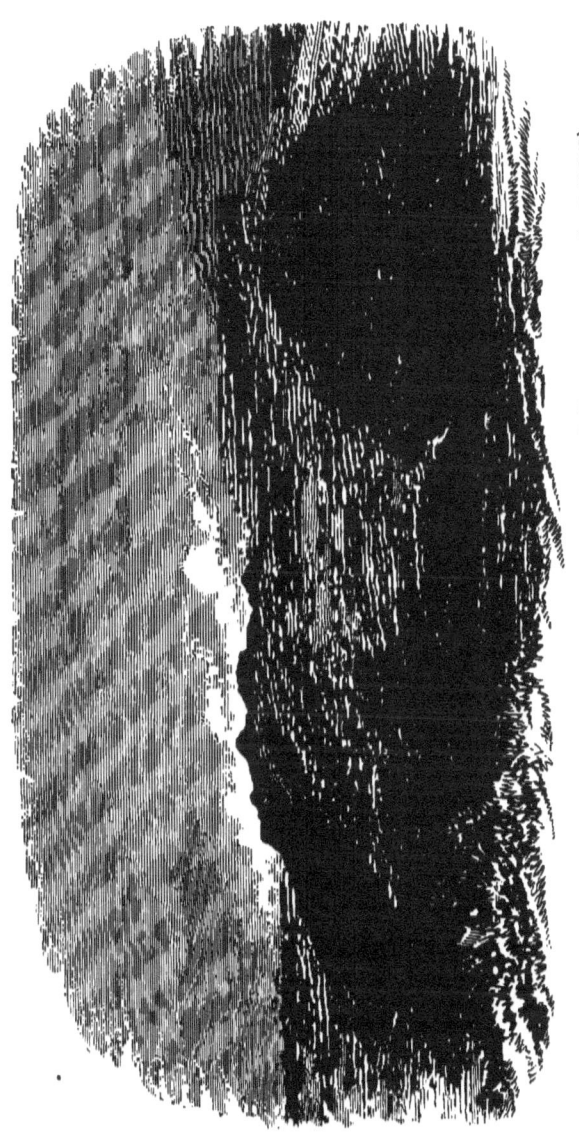

COCK'S TOR—VIEW TAKEN FROM GREAT STAPLE TOR ON THE EAST (FEATURE FORMED BY AMPHIBOLITE).

PLATE V.

GREAT STAPLE TOR ON THE WESTERN MARGIN OF DARTMOOR (MURAL JOINTING IN GRANITE).

PLATE VI.

Fig. 1.

ELVAN QUARRY AT LOWER GRENOFEN.

Fig. 2.

BRENT TOR FROM THE NORTH.

Fig. 3.

BRENT TOR FROM SMEAR RIDGE.

Fig. 4.

QUARRY AT GRESTON RIDGE.

·PART II.

MICROSCOPIC EXAMINATION OF THE ROCKS.*

No. 1. *Brent Tor* (north side).—A grey aphanitic rock containing irregularly-shaped patches, which on a polished surface have a deep reddish brown colour, and under a strong lens appear finely vesicular. These patches are much harder than the surrounding matrix, and seem to approximate to Eisenkiesel.

Under the microscope, by transmitted light, the red patches appear opaque. They are highly vesicular, and probably represent fragments of pumice which have undergone alteration, so that the fibrous structure is no longer visible. In some parts of the section irregular strings of dark opaque matter traverse the matrix and frequently follow the same direction, thus giving the appearance of fluxion structure ; this appearance may in part represent filamentous portions of pumice fragments cut parallel with the direction in which the filaments run. The matrix shows under a magnifying power of about 70 diameters great numbers of very minute doubly-refracting microliths and granules. It is seemingly a devitrified magma (the result being felsitic matter), but it is scarcely safe to express this opinion very decidedly. The microliths lie in all directions, and do not form definite streams. A little calcspar is here and there visible. The rock may be regarded as a pumice breccia. Fig. 2, Pl. VII., represents portion of a section of this rock magnified 25 diameters. The dark patch occupying the southern portion of the field is part of a scoriaceous fragment.

No. 2. *Brent Tor* (south side near the foot of the Tor).—A greenish-grey or buff-coloured rock, with dark grey imbedded fragments, and red stains from peroxidation of ferruginous matter. Polished surface shows minute greyish-white flecks and large spots more or less angular in form and of a dark green or black colour.

Under the microscope the rock appears to consist of a devitrified magma, now felsitic, which shows fluxion texture and contains fragments of vesicular rock (probably pumice). Quartz occurs in granules, globules, strings and nests, and is probably a secondary product, occupying the place of some mineral which has decomposed. The devitrification of the magma seems in part to be due to the development of doubly refracting colourless microliths. The rock appears to be essentially a pitchstone or hyaline rhyolite (devitrified) containing fragments of a scoriaceous rock (? pumice). It might be called a rhyolitic breccia. Fig. 1, Pl. VII., represents portion of a section of this rock as seen by ordinary transmitted light under a magnifying power of 25 diameters. The dark portion on the south-west edge of the field is part of one of the scoriaceous fragments, while the little irregular black flecks visible in other

---

* The numbers in the margin refer to the sections of these rocks which are deposited in the Museum of Practical Geology, Jermyn Street.

parts of the drawing doubtless represent other minute fragments of similar material included in the section. The white portions are composed of granules and crystals of quartz, among which globular forms are visible in places, but are not shown in the drawing. They sometimes exhibit, very imperfectly, the interference cross which characterises the polarization of radiate aggregates. In the coloured plate the pale yellow part represents the devitrified magma, the fluxion texture being shown by the deeper yellow markings. This drawing is a faithful rendering of the actual section, and I have taken pains neither to exaggerate nor to suppress any of the markings.

No. 3. *Brent Tor* (north-west side at the foot of the tor).—The specimen from which the section was cut is greyish in colour and consists of what appear to be angular fragments, ranging up to about an inch in diameter, and resembling fragments of slate imbedded in a paler grey matrix, suggesting at once that the rock is a breccia. On the polished surface it looks rather like a dark green serpentine variegated with whitish speckles and deep red ferruginous spots.

Under the microscope the section becomes dark between crossed nicols, the dark ground being, however, studded with innumerable doubly refracting microliths which observe no definite direction and are no doubt of secondary origin. These microliths represent far-advanced devitrification of what was, I believe, an originally glassy magma. Rounded greenish, feebly-dichroic granules of a mineral which is apparently augite and others which are colourless and singly-refracting are sparsely scattered through the section, which includes many patches of what also seems a devitrified glass but contains no microliths, only doubly-refracting granules. Some dark opaque matter, probably limonite, is also present in places. The microlithic patches in the section correspond with the dark slaty-looking patches in the hand specimen. Some of them show indications of a vesicular structure, the vesicles being occasionally bordered by dark margins, fragments of much altered minerals and scoriaceous dust, together with minute crystals, which in some cases may represent magnetite or pyrites, and in others a decomposed condition of minute scales of mica. Fluxion texture is clearly discernible in some of the microlithic patches. The rock is essentially a devitrified rhyolite.

No. 4. *Brent Tor* (north side).—A dark grey rock resembling a highly scoriaceous lava.

Under the microscope the section of this rock shows a quantity of opaque, finely-vesicular scoria, similar to that described as occurring in the preceding specimens (and which it was suggested might represent pumice). Here and there it seems that the rock is composed of fragments of this material. The vesicles are filled with siliceous matter, which has a very finely granular appearance when seen between crossed nicols, and similar substance surrounds the scoriaceous fragments. It is quite possible that the fragmentary aspect of this rock under the microscope may be deceptive, since the interspaces between the seeming fragments of scoria may merely represent sections of large, irregularly shaped vesicles which have like the smaller ones been filled up with siliceous matter. If the latter supposition be correct, and I am disposed to think that it is, we must regard this rock as a scoriaceous lava, while if preference be given to the former hypothesis, it can only be considered as an agglomerate of scoriaceous fragments and dust held together by a siliceous cement, and if so, this rock and No. 1, which is also from the northern side of the Tor, may be looked upon as identical. Under any circumstances they bear microscopically a very close relation to one another, although the general aspect of the hand specimens is very dissimilar, No. 1 being compact, while No. 4 has a rugged, scoriaceous appearance, highly suggestive of a lava.

No. 5. *Brent Tor* (north side).—The hand specimen is of very compact texture and of a deep reddish brown or Venetian-red colour, with small dark specks; this is best seen where the specimen has been cut.

Under the microscope the section is seen to contain numerous little prisms of felspar which at times show indications of twinning and are no doubt triclinic. With the exception of these prisms little can be made out beyond the fact that the surrounding matter displays a slightly granulated character . between crossed nicols, the light not being totally extinguished, while little clear doubly refracting microliths occur in it in considerable numbers. Dark matter is also present, some of which may be pseudomorphous after augite or other minerals, but no definite augite forms are to be discerned. Quartz occurs pseudomorphous, apparently after felspars and other minerals and also filling vesicular cavities. The rock is no doubt a decomposed basalt-lava.

No. 6. *Brent Tor* (north-west side at the foot of the Tor). — In general appearance this closely resembles the preceding. Under the microscope it is seen to be also a basalt with an originally glassy magma, which is now devitrified by the development of clear doubly refracting microliths. The section shows some irregularly shaped patches of brownish-yellow glass which contain comparatively few microliths, and with the exception of these bodies appear dark between crossed Nicols. There are indications of fluxion texture in the section. Any pyroxenic minerals which the rock may have originally contained are now decomposed, and the little prisms of felspar have also undergone considerable change. Although the evidence for pyroxene is unsatisfactory both in this and in No. 5, I have little doubt that both rocks represent altered conditions of what were once basalts with glassy magmas.

This is probably the rock described by Sir Henry De la Beche as a greenstone. In the absence of microscopic examination one might also feel disposed to apply the equally vague term aphanite to it. Pl. VIII. shows part of a section of this rock magnified 55 diameters, as seen by ordinary transmitted light.

No. 7. *Brent Tor* (south side at the foot of the Tor).— This is portion of a rounded, imbedded nodule, which, so far as can be judged from microscopic appearance, is allied to basalt. The section when magnified shows great numbers of rounded vesicles and more or less polygonal patches, all of which are filled with quartz or chalcedony; the latter forms, however, no doubt represent spaces once occupied by crystals which were normal constituents of the rock. Another imbedded nodule from the same locality shows larger vesicles, some of which are filled with a zeolitic mineral and others with quartz.

No. 8. From grubbing in roadside, turning to *West Langstone.* —This is a pale brownish-grey rock, and in the hand specimens shows little dark green patches, mostly rounded or lenticular, and moderate sized crystals, and fragments of crystals which once were felspars, but have now undergone so much change that they are easily scratched with the knife blade. They are greyish-white and form light blotches in the matrix. The dark-green patches are greenearth, or some closely allied mineral, and may have resulted from the alteration of fragments of some form of pyroxene; but in most cases I think they represent the filling up of vesicles.

A section of this rock when seen under the microscope shows a fine meshwork of felspar crystals considerably altered, and in general appearance it resembles a decomposed basalt, containing rounded fragments of larger felspar crystals, often with a peculiar zone of darker matter constituting their margins, suggestive of fritted or fused surfaces, and amygdaloidal patches of green earth or some allied mineral. This rock, which is definitely bedded, is overlaid by slaty beds containing manganese oxides and a little pyrites, in which is a layer of micro-crystalline pink-coloured quartzite, locally called capel (No. 8a), which contains a little pyrites and manganese oxides.

In the manganese mines in the neighbourhood capel often overlies the deposit of ore. It was formerly removed by means of the pick, but the process was a tedious and expensive one and the capel is now usually blasted.

The exposure from which the above specimens were taken occurs outside the greenstone boundaries marked on Sir Henry De la Beche's map, and is situated on the right-hand side of the main road between East and West Langstone.

No. 8b. Quartzite also occurs at Week (spelt on the ordnance map Wick), near Brent Tor. It is seen under the microscope to be rather more coarsely micro-crystalline than the capel at West Langstone, and contains minute crystals of decomposed pyrites.

No. 9. *Churlhanger*, near Lamerton.—The mode of occurrence of this rock has been already described (pp. 23 and 24). Its real character, so far as its origin is concerned, is by no means evident. It has a schistose structure and is of a greenish colour. It is profusely speckled with little amygdaloids of calcspar, which when weathered out give the rock quite a vesicular character. It may

be a fine ash, the volcanic dust having been comparatively un-
mixed with coarse fragments, but microscopic evidence upon this
point is unsatisfactory, and it is hard to conceive how an ash
deposit could become so vesicular.* It is also possible that it is
a much altered lava, but this assumption again receives but little
support from microscopic examination, while the schistose character of the rock rather militates against the supposition. The
vesicles are in many cases fringed with a green mineral which is
apparently chlorite, while the whole of the rock, except the calcspar
amygdaloids, when seen by transmitted light, partakes of the
same greenish tinge, variegated by dark opaque specks of irregular
and ill-defined form (Fig. 6, Pl. IX.). The rock is essentially an
amygdaloidal schist, and is almost identical in appearance with
some of the amygdaloidal Schalstein of Nassau. I have estimated
the carbonic anhydride contained in the Churlhanger rock from
an average sample, and find that it amounts to 15·5 per cent.
This gives a per-centage of 35·227 of calcspar, a result which
corresponds very closely with the estimate which might be roughly
formed from an examination of the microscopic section. After
deducting for the loss of carbonic anhydride and hygroscopic water
the loss on ignition is found to be 2 per cent., which represents
the combined water. This is probably derived from the green
mineral, and if we deduct 35 per cent. for the calcspar contained
in the rock, and take the remainder to consist of about one third
chlorite, the 2 per cent of combined water would suffice to make
good the assumption that the green matter is chlorite or some
allied mineral. After digesting the powdered rock in hydrochloric acid, and separating the alumina, iron, and lime, the solution
gives the usual reaction for magnesia.

The rock fuses to a dark-greenish, feebly-magnetic slag with a
thin vitreous crust.

No. 10. *Hardwick Quarry, Kilworthy,* near Brent Tor.—A
greenish-grey schistose rock; in places vesicular, in other parts
containing numerous imbedded fragments of a finely-vesicular
rock, which even in the thinnest sections transmit no light
except through the vesicles, which are filled either with silica
or with matter resembling the matrix in which the fragments
lie. The following sketch (Fig. 9), indicates the irregular

Fig. 9.

forms and disposition of some of these fragments as seen on a
smoothly cut surface (natural size). Some of the imbedded

* Although the amygdaloidal character of some Schalstein is mentioned in
Zirkel's Lehrbuch der Petrographie, yet he in no way attempts to account for
vesicular structure in schistose rocks when describing the origin and mode of
formation of amygdaloids.

fragments are probably of considerable size, judging from the cavities (C, C, C, C,) shown in the annexed rough sketch (Fig. 10) of a loose specimen, apparently an exceptional one,

Fig. 10.

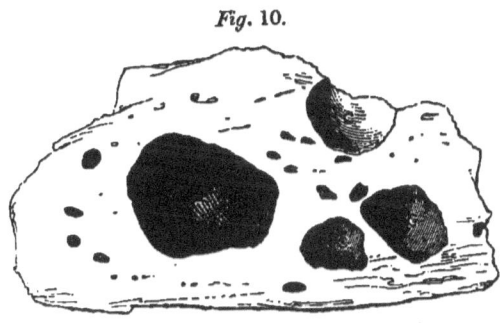

picked up in the quarry, since their dimensions rather negative the supposition that they are air or gas vesicles or steam pores, although numerous little vesicles of this kind surround them. They are mostly lined with a crust of pyrolusite or wad, but in one instance a recently broken lump contained in this specimen consists of soft white earthy matter like kaolin. The rock is spotted in places with stains of peroxide of iron. Under the microscope, sections from this quarry show well-marked schistose texture, closely resembling the fluxion structure so common in the adjacent rocks of Brent Tor itself, and suggesting the idea that *some* of the latter may merely represent an altered condition of beds similar to these schists, or what seems more unlikely, that the schistose beds are greatly altered lavas. The appearance of fluxion structure in these schists is probably deceptive, and represents the result of pressure which has compelled loosely coherent particles to adjust themselves in directions varying with the different unequal conditions of pressure and resistance caused by the harder, irregularly-shaped fragments which lie imbedded in the rock. The schist itself is filled with minute doubly-refracting microliths which appear to follow no definite directions, and have probably been developed since the rock was formed. Numerous little whitish granules, almost opaque, are also to be seen under the microscope. Roughly speaking the rock may be called an ash, and it is no doubt mixed to some extent with ordinary sedimentary matter. This stone is used in the construction of Kelly College at Tavistock. As a building stone it is likely to suffer more from disintegration by frost and damp, than from decomposition.

No. 11. *Langford Farm*, on road from Tavistock to Milton Abbot.—Bluish-grey slaty rock, irregularly fissile and schistose in places, the divisional planes coated with little scales of chlorite. The rock contains a few well-marked vesicular bands, which a little way in from the weathered surface become amygdaloidol, the vesicles being filled with calcspar, and the amygdaloidal bands thus bearing a close resemblance to the rock at Churlhanger. Regarded *en masse* the deposit no doubt represents a considerable thickness of ashy matter, mixed probably with ordinary sediment, once mud, now shale. I have not examined this rock microscopically, but it doubtless corresponds very closely with that from Churlhanger, already described, so far at least as the vesicular portions are concerned, while some of the more compact parts probably resemble the rock occurring at Kilworthy. The fissile structure corresponds approximately with the direction of the vesicular bands; the vesicles in these bands do not appear, as a rule, to be much compressed.

No. 12. *Brazen Tor*, near Petertavy (exposure mapped as greenstone).—A dark close-grained crystalline rock occurring close to the granite of Dartmoor, and in contact with it. The small imperfectly formed hornblende crystals which mainly compose the rock have a somewhat fibrous or silky appearance; the other components are quartz and a little finely disseminated pyrites. In thin section, when seen under the microscope, some of the hornblende crystals are strongly dichroic, others only feebly so, but the latter mostly lie in a different direction and are spiculæ rather than crystals. They are arranged in patches or streams in which they rest side by side in definite directions; the streams are sometimes slightly tortuous, so that they occasionally appear to have a wavy, fibrous texture. In Plate X. (Fig. 1) a drawing is given, showing portion of a thin section of this rock as seen by ordinary transmitted light under a magnifying power of 55 diameters. The irregularly shaped black patches represent pyrites, and the yellowish mineral hornblende. The rock is Amphibolite (hornblende rock).

No. 12a. Close to the beck near Waspworthy, near Brazen Tor.— This is another patch of rock mapped in as greenstone. It is identical with the preceding, both in its mode of occurrence and in its mineral composition. The rocks from these two localities appear to differ very slightly in texture.

No. 13. *Cock's Tor.*—An iron-grey or dark greenish-grey coarsely crystalline rock. It is in great part composed of a pyroxenic mineral, the imperfectly developed crystals of which have a somewhat bronze-like or metalloidal lustre. By Sir Henry de la Beche this mineral was thought to be hypersthene. Professor A. Rénard, of Louvain, who has kindly examined it for me, believes it to be diallage; while Professor Rosenbusch, of Strassburg, to whom specimens were forwarded by Professor Rénard, takes it to be augite. In presence of two such good authorities I feel some diffidence in expressing any strong opinion upon this subject; still I am inclined to think with Professor Rénard that, from the metalloid lustre, and from the difference in intensity of the two cleavages in some of the crystals, they are probably diallage. The green dichroic matter which is present apparently results from the alteration of some of the pyroxenic mineral into viridite. Titaniferous iron and felspars, both of which have undergone considerable alteration, are the other components. The nature of the felspars is not evident, owing to the change which they have suffered, but I am inclined to think that they are only plagioclastic in part, one or two of the crystals exhibiting indications of only a single twinning plane. Where alteration of the titaniferous iron has taken place, a white substance is developed whose precise chemical nature is yet unknown. In a paper, "Some Results of a Microscopical Study of the Belgian Plutonic Rocks," read before the Royal Microscopical Society (April 5, 1876), Professor Rénard speaks of this alteration of the Ilmenite in the Gabbro of Hozémont in the following terms:—"The sections of this titanic iron are surrounded and covered in some cases with coatings of an opaline substance, perfectly homogeneous, which seems a result of the decomposition of ilmenite. The first stage of this decomposition is represented by the appearance of whitish veins running through the mineral; a second stage exhibits it enclosed in the opaline substance; finally, the metamorphosis can be pushed so far that nothing more is visible except a few black specks. Its chemical composition has not been determined, but we have ascertained that it is unalterable by the action of hydrochloric acid, and therefore it is not carbonate of iron, as has been taught by some. We are, however, persuaded that the opinion of Gümbel, who admits that it is not a decomposition product, cannot be sustained."[*] Fig. 4, Plate X., shows this alteration of titaniferous iron in the Cock's Tor rock. The drawing was made partly by reflected and partly by transmitted light simultaneously, so that the alteration product appears of a greyish or greenish-grey colour instead of white, as it does by reflected illumination only. Believing with Professor Rénard that the pyroxenic mineral in this rock is diallage, I regard the rock as Gabbro, and have marked it as such on the small map prefixed to this Memoir.[†] Should the mineral, however, be really augite, as Professor Rosenbusch is disposed to think, the name would probably have to be changed to diabase, but even before this point could be satisfactorily settled it would have to be shown, according to Dathe,[‡] that the plagioclase present was oligoclase.

No. 14. *Indescombe*, near Tavistock.—A greenish-grey crystalline rock containing dark-green, imperfectly-developed crystals of a pyroxenic mineral. The rock weathers with a light-greyish crust, and to the naked eye closely resembles diabase. The component minerals appear under the microscope to be decomposed felspars (plagioclase), pyroxene (probably augite), some magnetite, and a little quartz. It is a rock closely allied to, if not basalt. If chlorite were present it might be regarded as diabase. The alteration is mainly restricted to the felspars and the magnetite.

---

[*] The Monthly Microscopical Journal and Transactions of the Royal Microscopical Society, Vol. XV., page 217.

[†] Since this was written, Mr. S. Allport, in a paper "On the Metamorphic Rocks surrounding the Land's End Mass of Granite." Quart. Journ. Geol. Soc., Lond., Vol. XXXII., p. 421, has described the rocks of Cock's Tor and Smear Ridge as "altered gabbro or dolerite." (See page 49.)

[‡] Zeitschrift d. deutschen geolog. Gesellsch, Vol. XXVI., page 1.

No. 15. *Cottage Inn,* on main road from Tavistock to Mary-tavy (about half a mile east of Indescombe).—This rock is rather more coarsely crystalline than that occurring at Indescombe, but bears a strong resemblance to it both in its colour and general aspect. The pyroxenic component of this rock has, however, a somewhat bronze-like lustre in places, and exhibits under the microscope cleavages of different intensity, similar to those which occur in diallage. Professors Rénard and Rosenbusch appear to differ in opinion upon this subject, the former authority believing he mineral to be diallage, while the latter considers that it is augite. If diallage, the rock is gabbro; if augite, it is basalt. I am disposed to think with Professor Rénard that it is gabbro. Still the rock at Indescombe, although very similar, appears from the cleavages in its pyroxenic component to more closely resemble a basalt. I have nevertheless no doubt whatever that the exposures at Indescombe and at this locality are connected, and that the rocks, if they differ at all, pass by gradation into one another.

Fig. 3, Plate X., shows the cleavages of unequal intensity in the pyroxenic mineral, which coupled with its metalloid lustre favour the supposition that it is diallage, and consequently that the rock is gabbro.

No. 16. A rock occurring near Ridge on the opposite side of the Tavy corresponds very closely with the above in microscopic characters, but appears to contain more titaniferous iron, in great part converted into opacite. This exposure at Ridge is also doubtless connected with those at the Cottage Inn and Indescombe.

The following extract from a letter received from Professor Rénard, after he had examined the specimens and microscopic sections from Brazen Tor, Cock's Tor, and the Cottage Inn, cannot fail to interest those who may subsequently devote attention to these rocks.

The first paragraph is an extract from a letter which he received from Professor Rosenbusch :—" After a careful examination of the mineral I find that it belongs to the pyroxene group, but I do not think that it is diallage, since it fails to show the cleavage and the characteristic interpositions of that mineral. I am inclined to regard it simply as augite of the type which one meets with in the palæolithic diabases."

Professor Rénard adds :—" In the specimen of your rock, the minerals are not sufficiently well terminated to show their crystalline form. The cleavages exhibit a more or less metalloid aspect, which, to my mind, is more suggestive of diallage than augite. As to the lines of cleavage, such as they appear under the microscope, it is difficult to say to what mineral they belong. The absence of interpositions does not appear to me to prove that the mineral is not diallage."

With regard to the pyroxenic components of the rocks at Cock's Tor, Brazen Tor, &c. the following notes from the writings of Zirkel and Boricky may also prove interesting :—

Zirkel, in his Mikroscopische Beschaffenheit der Mineralien und

Gesteine, pp. 181–3, states that the very feeble dichroism of diallage indicates that it is allied in optical properties to augite and also renders it easy to be distinguished from hypersthene and hornblende, and he adds that the mineral in the so-called hypersthenites of Penig, Neurode, and the Isle of Skye, which was formerly described as hypersthene must now be regarded as diallage.

Boricky, in his "Petrographische Studien an den Melaphyr Gesteine Böhmens," page 19, describes a mineral, which occurs in some of the melaphyres of Bohemia, as an augite-like diallage. The sections of this mineral show, under the microscope, numerous parallel rectilinear fissures which he regards as the edges of delicate twin-like lamellar intergrowths, probably lying parallel to the basal planes of the crystals. In longitudinal sections of the crystals, these fissures are seen to cross the coarser cleavage planes at angles of 70° to 90° and sometimes occur only in one part of an individual augite section. The mineral shows no dichroism.

No. 17. *Lower Grenofen Quarry,* near Tavistock.—The chief components of this rock are orthoclase, twinned on the Carlsbad type, biotite, a little muscovite, and quartz.

Under the microscope the orthoclase crystals are seen in many cases to have undergone considerable change. A granulation or development of opaque-white flocculent matter setting in around the margins of the crystals. Some of the crystals are completely altered in this manner, so that they no longer exhibit any definite optical character. This physical change does not, however, appear to affect the hardness of the crystals, nor to impair the value of the stone for building purposes, for which it is well adapted. The orthoclase crystals range from half an inch in length downwards, giving a porphyritic character to the stone, which renders it suitable for ornamental purposes in architecture. The rock is essentially a fine-grained porphyritic granite with a little felsitic material. It may also be regarded as a quartz porphyry, although the porphyritic character is mainly due to the large development of the felspar crystals.

No. 18. *Shilla Mill Quarry* (light porphyritic variety).— Under the microscope the components are seen to be orthoclase, mainly decomposed, quartz, and magnesian mica. The matrix is felsitic, crystals of magnesian mica are in places seen to lie within the crystals of orthoclase, and also a few crystals of quartz. A re-entering angle in one of the porphyritic crystals of orthoclase is well marked in the section, but decomposition is far advanced. The twining is upon the usual Carlsbad type. Crystals of magnesian mica are also developed within this crystal. They appear to lie parallel to the clinodome. The angles of the quartz crystals are mostly rounded. The rock is a quartz porphyry or elvan. In the same section crystals of magnesian mica are to be seen lying within quartz crystals. From this it would appear that the mica was the mineral which first crystallized.

No. 19. *Shilla Mill Quarry* (dark compact variety).—In this section again magnesian mica lies within the quartz crystals. The felspar is in all cases converted into felsitic matter. In this section quartz, when in juxtaposition with magnesian mica, has asserted its form. In some instances quartz appears to be pseudomorphous after felspars.

No. 20. *Shilla Mill Quarry.*—In this section again the magnesian mica is developed within crystals of quartz. The matrix is felsitic. The felspar crystals are all decomposed, but the change which they have undergone is different from that which has taken place in the two sections just described. In the case of No. 18, the crystals appear white by reflected light, the change having apparently resulted from the conversion of the felspar into some substance allied to kaolin. In the section No. 19 the felspars are changed into felsitic matter, while in this section they seem to be replaced partly by a micaceous mineral now greatly decomposed, and probably represented by limonite, and partly by felsitic matter.

Although the three sections of rocks just described are identical in their original composition, yet they differ greatly in their external aspect.

No. 21. Elvan-course in granite, *Gunnis Lake Company's Quarry, Hingston Down.*— A light-grey rock composed of orthoclase, a little plagioclase, brown and white micas, and quartz. The felspars have for the most part a granulated appearance when seen by polarized light under the microscope. The quartz contains some minute fluid cavities. The rock is essentially a fine-grained granite, and cannot be regarded as a true quartz-porphyry or elvan since there is no micro-felsitic magma.

No. 22. Top of *Brazen Tor*, near Petertavy.—This is a coarsely-crystalline granite, the chief components being orthoclase, plagioclase, biotite, and quartz, with segregations of schorl. The latter stand out in relief on the weathered surfaces of the rock, but seem to be capriciously distributed through it. The section examined shows some interesting fluid cavities, some of which in addition to bubbles contain minute cubes of rock salt. Some of the best examples are shown in Fig. 5, Pl. IX., as seen under a magnifying power of 400 diameters. The orthoclase crystals in this rock sometimes exceed 2 inches in diameter, rendering it very coarsely porphyritic, and this within 100 yards of the junction of the granite with the amphibolite. For the greater part of this distance, however, the rock is very fine-grained and of a pale pink tint, with large black segregations of schorl. In a section of this rock lines are visible in some of the plagioclase crystals cutting at right angles across the twin lamellæ,* others are altered by the development of a finely-granular structure. In some cases, although this granulation has extended over the entire crystal, the twin lamellæ are still to be recognised. Fig. 4, Pl. IX. shows a crystal of biotite lying within a quartz crystal in this section; while in other instances the biotite lies partially

---

* In a paper read before the Geological Society, May 12, 1875, the author pointed out the existence of similar transverse bands in plagioclase from the Gabbro of Volpersdorf in Silesia.

enveloped within crystals of felspar. Little crystals of quartz are also to be seen enveloped by felspar. From these facts it would appear that the felspar was one of the last minerals to crystallize, and the slowness of this process may in some measure account for the large size which some of these crystals attain.

No. 23. *Brazen Tor* (near the contact of the granite with amphibolite). This is the fine-grained pale-pink rock with black segregations of schorl, just alluded to as lying between the top of Brazen Tor, where the coarsely-crystallized porphyritic granite occurs, and the amphibolite. The section when examined under the microscope shows no trace of mica, while the hand specimens appear to be equally deficient in this mineral. The rock is composed of plagioclase, orthoclase, quartz, and schorl. The orthoclase crystals are mostly Carlsbad twins. The schorl does not appear to be well crystallized, yet this mineral, as well as the quartz, is seen at times to encroach upon the areas belonging to the plagioclase crystals. A few minute singly-refracting granules occur in the section, which are probably garnets. As felspars and quartz are the principal mineral components of the rock and micas are absent, we are bound to regard this marginal portion of the granite as granulite or haplite. Fig. 2, Pl. IX., shows some of the schorl in a section of this rock, magnified 25 diameters and viewed by ordinary transmitted light. The blue and deep yellow portions of the drawing represent schorl, the white spaces quartz, and the pale-yellowish parts crystals, mostly of plagioclase which are somewhat altered.

No. 24. *Greston Bridge Quarry.* Slate below the lower flow of altered greenstone (? basalt).—This rock is an altered slate or Lydian stone veined with quartz. It is of a brownish-black colour, weathering light-brown, and has a splintery or subconchoidal fracture.

Under the microscope colourless microliths are visible, and they lie approximately in the direction which the longest axes of the component particles follow. This direction may be either that of cleavage or bedding, but probably the latter, as there is no distinct cleavage. The veins which traverse the slate have very sharp outlines, and were no doubt formed and filled after the rock had undergone most perfect consolidation. The veins are in part composed of crystals or granules of·quartz and partly by a finer crystalline and apparently microlithic substance in which quartz predominates.

In Fig. 2, Plate IX., a portion of the magnified section of this rock is delineated, showing a slight flexure of the direction of lamination in contact with one of the quartz veins; which implies considerable pressure and consequent molecular disturbance after the veins were filled.

No. 25. *Greston Bridge Quarry.* (Slate between the flows of altered greenstone).—This is a section of a similar slate to that just described. It shows well-marked lamination, and under the microscope a few colourless microliths are visible, but they do not invariably lie in the direction of the planes of lamination.

C. 31.   D

Neither in this nor in the preceding rock was there any stronger evidence of alteration at the bottom than at the top of the slate bed. Whatever has taken place, change physical or otherwise, it seems to have affected these little slate beds (only a few feet in thickness) equally throughout.

No. 26. Chert band in upper bed of altered greenstone. *Greston Bridge Quarry.*—In this section the only interesting features are the occasional imperfect development of quartz crystals, which in some cases have segregated in the vicinity of crystals of iron pyrites, the latter sometimes having the form of definite cubes. These chert bands are of a deep greenish-grey colour and seem to have been crushed and re-cemented. They run irregularly through the altered greenstone and are only a few inches in thickness.

No. 27. *Greston Bridge Quarry.*—Under the microscope very little can be made out concerning the original mineral composition of this rock. Elongated prisms which represent what once were felspars alone remain to attest its eruptive character, a supposition which becomes a certainty when the indurated nature of the associated slate beds is taken into consideration. The rock is mainly converted into serpentine. Calcspar fills up the joint-planes and is disseminated through the rock, yet it plays quite an unimportant part in its composition, since an average sample of the stone when ground smooth and immersed in acid underwent only very trifling erosion. A dark mineral is visible in the sections of this rock which is probably limonite, and which shows indications of being pseudomorphous after pyroxene or amphibole, and also after pyrites or magnetite but no well-developed forms are discernible. Some of the altered felspar crystals still seem to show traces of triclinic banding, but the appearances under polarized light are so hazy that it is unsafe to hazard any definite opinion about the minerals. In one section opaque acicular bodies are developed within nebulous areas which seemingly represent crystals, but the margins of the latter are ill-defined and they present no trustworthy optical characters. A drawing of portion of a magnified section of this very unsatisfactory rock is given in Fig. 1, Plate IX.

A section cut from a small specimen procured in the roadside on the north of the quarry shows numerous dark green spots about the tenth of an inch in diameter. These have possibly resulted from the alteration of olivine.

PART III.

DEDUCTIONS.

The deductions to be drawn from the foregoing observations serve to show that the views entertained by the late Sir Henry de la Beche are in the main correct, and represent a vast amount of truth derived simply from observations in the field. The doubt which he expresses relative to the boundary line between the Devonian deposits and the culm series, still leaves it an open question whether the eruptive rocks which lie around Brent Tor belong exclusively to the latter period or not. While if the different belts of ash marked on the map represent a repetition of the same beds, which appears highly likely, it precludes the supposition that they were deposited partly in Devonian and partly in carboniferous times. In all probability the boundary between the rocks of these two periods lies somewhat to the south of Tavistock. On the Survey map all the exposures of eruptive rocks are coloured as greenstone, but the " Report " upon the district shows clearly enough that they were not regarded as one and the same rock. Upon the small map prefixed to this Memoir an endeavour has been made to indicate some of the lithological differences which exist, and it is shown that the rocks coloured as one on the Survey map represent Amphibolites, Gabbros, Basalts, Pitchstones, and Schistose ashes or clastic rocks of a doubtful nature. One point is especially worthy of remark, namely, that some of the rocks which in part compose Brent Tor itself are apparently unrepresented in any other portion of the district, while another is that the Tor forms the highest point of ground for some considerable distance around. These are two very significant facts, and a right interpretation of them may do much to elucidate the geological history of this neighbourhood. After careful consideration of the question, it seems probable that Brent Tor as it now exists, only represents a portion of the series of volcanic ejectamenta and lava flows which have taken place, as suggested by Sir Henry de la Beche, at or near a crater. What has become of the remainder of these rocks, and to what is the preservation of this vestige of their former existence due? Brent Tor by no means represents a complete crater, for, if it did represent an entire structure of this kind, it is evident that outcrops of rock of a similar character would occur in the vicinity. Furthermore, since these rocks are much harder and less susceptible of weathering than those which surround them, it is evident that if this were the case any other outcrops of them which existed would form prominent features in the country. Since there is no indication of this, it is clear that Brent Tor represents the last trace of volcanic rocks of that particular character remaining in the district, and we have therefore only to deal with the two questions just enunciated. What has become of the remainder of these rocks and to what is the preservation of this little volcanic legacy due? It *might* be due to the Tor having occupied the lowest portion of a synclinal fold, while all the corresponding rocks rested at higher levels and were denuded, but the gentle undulations into which the surrounding beds are bent negatives such an hypothesis. As a last resource we are thrown back upon a fault, and it appears very probable that one *may* exist immediately to the east of the Tor.

I can adduce no evidence in support of this supposition other than that suggested by the foregoing remarks, and by the break in continuity of some of the schistose beds shown as volcanic ash on the Survey map. If, however, these beds represent corresponding beds east and west, the assumption that a fault exists requires little additional proof, while if it be inferred that some of those belts of ash which observe an annular disposition around the eastern and southern aspects of the Tor, dip inwards and form a synclinal fold, it follows that they are superficial representatives of one and the same bed. If this much be granted, it is easy to account for the diminished distance between their assumed continuations on the north and east of the Tor, for if the fault be considered to run in a N.N.W. direction, and the downcast to have taken place on the western side, it is manifest that after denudation the chord of the eastern synclinal arc will be less than that of the more depressed western syncline. It has, however, yet to be demonstrated that the ash belts lying east and west of the assumed line of fault do really represent continuations of the same beds; this can only be shown, so far as I know, by better and more satisfactory exposures of these rocks than those now visible; moreover, the strip of rock between Bowdon and West Langstone, and probably that between Shallor and Lower Chillaton, dip away from Brent Tor. Many of the changes in dip may, however, be due to very slight undulations. We have then a series of ash beds and clastic rocks girding Brent Tor which are nearly related to it in date of formation, both being of Carboniferous age and occupying closely succeeding horizons. Brent Tor would have been swept away ages ago had it not been nested at a lower level than the now demolished remains of the crater and series of lava flows and agglomerates of which it once formed a part. The balance of evidence, such as there is, tends to show that it owes it preservation to a fault, a fault unmapped and the existence of which is as yet purely a matter of inference.

The masses of gabbro extending between Crayston and White Tor, Wheal Friendship and Smear Ridge, are no doubt connected with one another at some depth beneath the surface, and it is probable that the small patches occurring in contact with the granite at Waspworthy and Brazen Tor and the strip extending from Cock's Tor to Indescombe, are offshoots from the same deep-seated mass. These may even be of later date than the granite of Dartmoor, and so far as evidence goes, there is nothing to show that they are in any way connected with the volcanic series of Brent Tor and its surrounding ash beds.

Upon the rocks at Greston Bridge and Dunterton it is very difficult to offer any opinion, since they are so much decomposed (and in great part converted into serpentine) that it is scarcely safe to hazard any definite suggestion as to their original composition. They *may* have been basalts. One thing, however, is certain, namely, that they represent lava flows, separated by a short period when ordinary sedimentary deposits were formed. Lack of time precluded the possibility of making more extended observations upon those portions of the ash and trap belts which are left blank upon the map which accompanies this Memoir; but in most instances, where they were examined, scarcely any evidence could be procured other than that afforded by road metal; the land was for the most part under cultivation, the few exposures which were visible were not trustworthy, and, unless better evidence were to be procured when the Survey map was made, it is clear that some of the boundary lines must have been drawn from the superficial features of the ground. Still, where reliable outcrops occur and where I have had an opportunity of verifying the work done by the founder of our survey nearly forty years ago, there seems no reason whatever to doubt its accuracy, nor the clearness of perception with which boundary lines were drawn and rocks mapped-in, even where the evidence was scant and the record hard to be deciphered.

The eruptive rocks of Brent Tor and its immediate neighbourhood apparently bear no relation to those in the neighbourhood of Exeter, on the eastern side of Dartmoor. The latter may represent a continuation of the volcanic action which took place in Carboniferous times, but that they are of much later date there can be no question; for although in places they are very similar to the rocks of the Tavistock district, the sedimentary deposits with which they are associated belong to a later geological period. In considering the history of this district it is important to bear in mind the stratigraphical relations of the sedimentary rocks, the displacements, vertical and lateral, which they have undergone, and the enormous amount of material which has been removed by denudation. The Devonian and the Culm series apparently rest conformably on one another in this area, and in the absence of any strong palæontological evidence to the contrary we may regard the two series as representing one uninterrupted period of deposition of sediment under varying bathymetrical conditions; but the local paucity of fossils greatly diminishes the chance of finding organic evidence in connexion with such changes.

No serious stratigraphical disturbance in this small area marks the period when these old marine deposits took place, although, from the evidence furnished by the eruptive rocks, it is clear that volcanic action was going on during the deposition of the lower portion of the Culm series. It may perhaps be due to the opening of small safety valves that no great flexures were developed in the strata through which these rocks were extruded and with which they are interbedded, and we may with some show of reason believe that this state of things continued until nearly the whole of the Culm rocks were deposited. At a subsequent period upheaval ensued, and it was probably then that the Devonian and Culm series underwent some lateral pressure which threw them into the major synclinal fold, which now brings the Devonian rocks to the north and south of the Culm area, and at the same time produced those minor undulations which are obscurely indicated by not very numerous and often not very trustworthy dips.

After or during upheaval marine denudation commenced its work along the margins of the emerging land, but what time elapsed before this important elevation of the district took place it is hard to say. It probably occurred towards the close of the Carboniferous period or in Permian times, and it was most likely during this upheaval that the intrusive granitic masses which constitute Dartmoor, Hingston Down, &c., forced their way into the Culm series, in places uptilting the surrounding beds, as pointed out by Sir Henry De la Beche.

The occurrence of granitic debris in the Triassic rocks seems to indicate that rocks of granitic character were denuded before or during the deposition of the Triassic beds; yet although the latter have also in great part been removed by similar agencies, it is probable that granitoid rocks connected with those of Dartmoor and some of the other adjacent granite areas may have formed islands in comparatively early Mesozoic times Through subsequent changes of level coupled with periods of marine and of subaerial denudation the country has assumed its present surface configuration, but it seems likely that the harder eruptive rocks of this district have at all times resisted the destructive action of the sea as well as that of rain and rivers, far more than the softer rocks which surrounded them ; and consequently it is only reasonable to assume that at whatever period marine denudation took place after these hard rocks were once laid bare, they would invariably be the spots which offered the greatest resistance when upheaval and denudation of subsequent deposits ensued : and that at all events, if the process of upheaval were a slow one, they formed small islands, and may even have existed as such within comparatively late geological times.

Tinstone and copper pyrites constitute the chief metalliferous lodes of importance in and around these granitic masses. Manganese ores, arsenical pyrites, and hematite also occur in their neighbourhood, but the former are more plentiful at a distance from the granite, the presence of which also seems to be inimical to the development of lead ores.

Since this work was written a paper was read by Mr. Samuel Allport before the Geological Society (June 21, 1876) "On the Metamorphic Rocks surrounding the Land's-End Mass of Granite."

In this paper the author compares some of the rocks of Brent Tor and the Tavistock neighbourhood with those which occur in the Penzance district. The following extracts from the paper confirm some of the statements made in this Memoir, but Mr. Allport does not appear to have detected the rhyolitic character of some of the rocks which compose Brent Tor.

"At Peter Tavy the rock forming Smear Ridge and its prolongation to Cock's Tor, near the granite, is an altered gabbro or dolerite, some specimens having a coarsely crystalline texture, while others are fine-grained; both varieties are rather less altered than the Penzance rocks, and are, therefore, important for comparison.

"In the coarsely crystalline variety the alteration is of precisely the same character as that described in the Tolcarn rock; the diallage has been partially converted into hornblende, and this substance also fills veins and former cavities; the felspar, though highly altered may be readily recognised; and there are the same pseudomorphs after magnetite.

"In a fine-grained specimen of the same mass the pyroxene (apparently augite) is very well preserved, but the felspar is completely decomposed, being represented by pale green pseudomorphs; the same substance also fills cavities, several of which contain groups of radiating blades of tremolite.

"Brentor, four miles north of Tavistock, presents many of the features of a volcanic mass, being chiefly composed of purple bedded ash, together with scoriaceous and compact trap of a greenish-grey colour.

"The latter is an altered basalt, in which the augite occurs in small well-formed crystals; it has suffered little or no alteration, while the felspar is converted into pale green pseudomorphs.

"The mass of 'greenstone,' half a mile north-east of Brentor, is chiefly ash of the same character as that of Brentor itself."—Quart. Journ. Geol. Soc., Lond., vol. XXXII., p. 421.

The last statement, here quoted from Mr. Allport's paper, seems to me to refer to an exposure of schistose rock visible in a narrow lane near Brent Tor. The rock did not impress me with the idea that it was identical with any of those which actually constitute the Tor; and, although I collected a few specimens, only one of these has yet been examined microscopically. This is a vesicular basalt, with a matrix which, in part, is singly refracting. The augite is mainly decomposed. If Mr. Allport's conclusion be a correct one, it may serve to indicate the down-throw-side of the problematical fault, the existence of which I have endeavoured to demonstrate. I should, however, have anticipated that, where rocks identical with those of Brent Tor occurred, a more marked feature would have been developed.

POSTSCRIPT.

During the revision of this Memoir I have read an interesting paper by Dr. Harvey B. oll, " On the Older Rocks of South

Devon and East Cornwall," published in the Quarterly Journal of the Geological Society for November 1868. In this paper Dr. Holl states that there are numerous faults in the vicinity of Brent Tor, one of them crossing the River Lyd, south-west of Coryton, and ranging by Monkstone to the west of the Tor. " Another skirts the north-east side of the volcanic rocks of the Tor, extending from near Monkstone to South Brent Tor. These faults carry the country further to the south or rather south-east, and reverse the dips along the valley of the Tavy, where the beds rise to an anticlinal axis, which crosses the railway in a N.N.E. and S.S.W. direction, midway between Ford Gate and the Marytavy Railway Station."

Although Dr. Holl does not appear to have investigated the important bearing which these faults have upon the question of the preservation of Brent Tor itself, I am glad to find that, his statements, so far as the directions of these faults are concerned, coincide approximately, and in one instance perfectly, with the conclusions at which I had independently arrived.

APPENDIX.

LIST OF MINERALS FROM THE NEIGHBOURHOOD OF TAVISTOCK.

Actinolite.*	Fluor.
Anatase.	Galena.
Augite.*	Garnet.
Axinite.*	Jasper.*
Bismuthine.	Kupfernickel.
Bitumen.	Limonite.
Blende.	Manganite.
Brookite.	Marcasite.
Calcite.	Mispickel.
Chalcotrichite.	Olivenite.
Chalybite.	Psilomelane.*
Childrenite.	Pyrites.
Chlorite.	Pyrolusite.*
Clinoclase.	Quartz.
Copper, Native.	Rhodonite.
Copper Pyrites.	Scheelite.
Copper-uranite.	Silver, Native.
Cuprite.	Titaniferous Iron.
Diallage.	Vivianite.
Erythrine.	Wolfram.
Fahlerz.	

---

* These minerals have been stated to occur at Brent Tor. The immediate neighbourhood is probably meant, not merely the tor itself.

INDEX.

A.

B.

C.

G.

H.

I.

J.

K.

L.

M.

N.

O.

P.

V.

W.

Z.

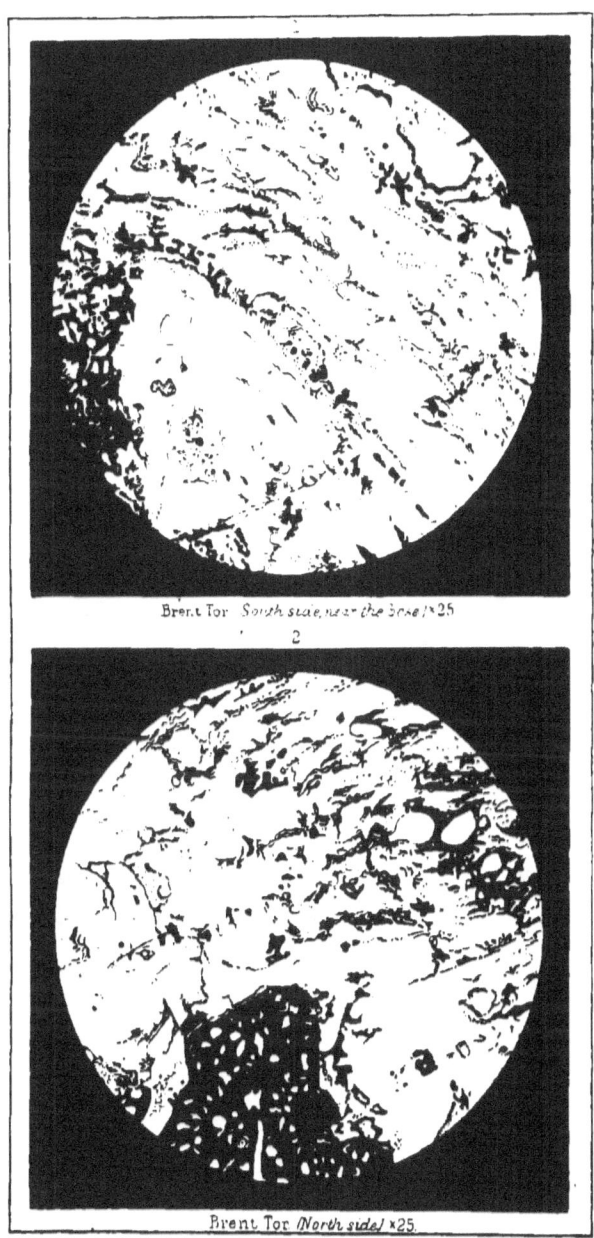

Brent Tor *South side, near the base* ×25

2

Brent Tor *(North side)* ×25

Frank Rutley del.

Vincent Brooks Day & Son. Lith

BASALT ICE No. 1

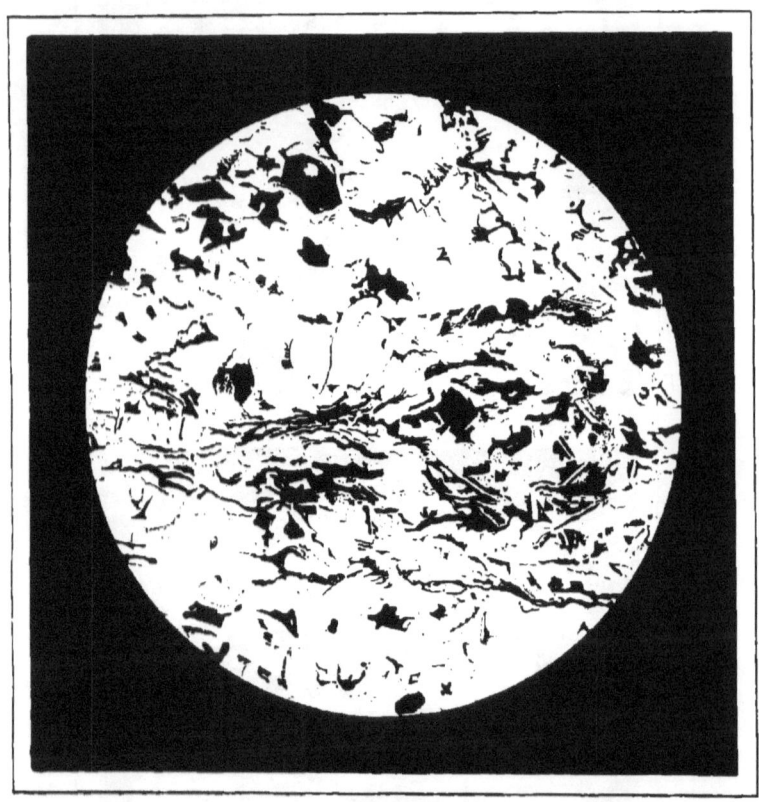

Basalt with glassy magma mainly devitrified.

Frank Rutley del.

Vincent Brooks, Day & Son Lith

Plate X

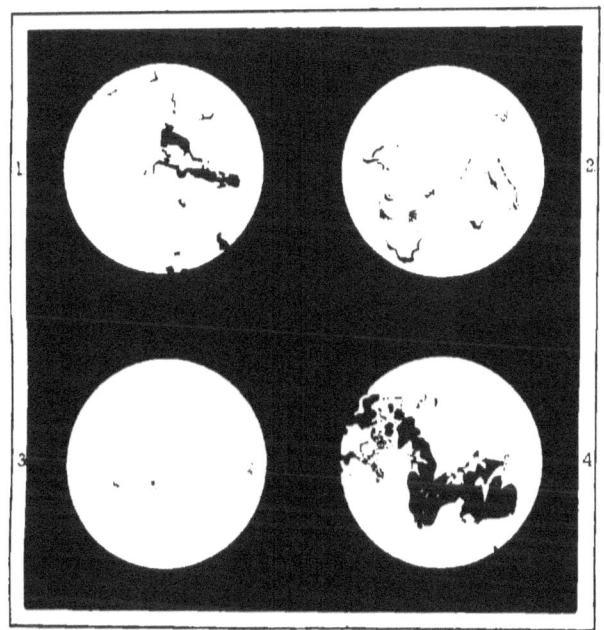

Nº1 Hornblende, Pyrites and Quartz in Amphibolite, Brazen Tor, Devon
„ 2 Schorl in the Granulitic margin of the Granite in contact with Nº 1
„ 3 Pyroxene (probably Dialliage) in Gabbro, Cottage Inn Main Road to Marytavy
4 Titaniferous Iron partly altered (Gabbro) Cocks Tor, near Tavistock

Nº1, ×55. Nºˢ 2,3 and 4 ×25. Ordinary transmitted light. On the surface of
Nº 4 a little light was also reflected simultaneously